S0-ADZ-077

Effect of IL-10 and anti-TGF-beta
antibodies on the morphology of
bone marrow stroma cultures
from
Interleukin-10
by
Jan E. DeVries and
Rene de Waal Malefyt
© RG Landes Co. 1995

MOLECULAR
BIOLOGY
INTELLIGENCE
UNIT

# G PROTEIN-COUPLED RECEPTORS

Tiina P. Iismaa, Ph.D.
Trevor J. Biden, Ph.D.
John Shine, Ph.D.

Garvan Institute of Medical Research
St. Vincent's Hospital
Sydney, Australia

R.G. LANDES COMPANY
AUSTIN

# MOLECULAR BIOLOGY INTELLIGENCE UNIT

## G PROTEIN-COUPLED RECEPTORS

R.G. LANDES COMPANY
Austin, Texas, U.S.A.

Submitted: October 1994
Published: January 1995

U.S. and Canada Copyright © 1995 R.G. Landes Company
All rights reserved. Printed in the U.S.A.

Please address all inquiries to the Publisher:
R.G. Landes Company, 909 Pine Street, Georgetown, Texas, U.S.A. 78626
or
P.O. Box 4858, Austin, Texas, U.S.A. 78765
Phone: 512/ 863 7762; FAX: 512/ 863 0081

U.S. and Canada ISBN 1-57059-058-3

International Copyright © 1995 Springer-Verlag, Heidelberg, Germany
All rights reserved.

International ISBN 3-540-58887-6

While the authors, editors and publisher believe that drug selection and dosage and the specifications and usage of equipment and devices, as set forth in this book, are in accord with current recommendations and practice at the time of publication, they make no warranty, expressed or implied, with respect to material described in this book. In view of the ongoing research, equipment development, changes in governmental regulations and the rapid accumulation of information relating to the biomedical sciences, the reader is urged to carefully review and evaluate the information provided herein.

### Library of Congress Cataloging-in-Publication Data

Iismaa, Tiina P.
    G protein-coupled receptors / Tiina P. Iismaa, Trevor J. Biden, John Shine.
        p. cm.—Molecular biology intelligence unit)
    Includes bibliographical references and index.
    ISBN 1-57059-058-3
    1. G proteins—Receptors. 2. Cell receptors. 3. Cellular signal transduction.
    I. Biden, Trevor J. II. Shine, John. III. Title. IV. Series.
QP552. G16I35 1995                                                    94-38448
574.19'245—dc20                                                       CIP

QP
552
.G16
I35
1995

# Publisher's Note

R.G. Landes Company publishes five book series: *Medical Intelligence Unit, Molecular Biology Intelligence Unit, Neuroscience Intelligence Unit, Tissue Engineering Intelligence Unit* and *Biotechnology Intelligence Unit.* The authors of our books are acknowledged leaders in their fields and the topics are unique. Almost without exception, no other similar books exist on these topics.

Our goal is to publish books in important and rapidly changing areas of medicine for sophisticated researchers and clinicians. To achieve this goal, we have accelerated our publishing program to conform to the fast pace in which information grows in biomedical science. Most of our books are published within 90 to 120 days of receipt of the manuscript. We would like to thank our readers for their continuing interest and welcome any comments or suggestions they may have for future books.

Deborah Muir Molsberry
Publications Director
R.G. Landes Company

# CONTENTS

# ABBREVIATIONS

| | | |
|---|---|---|
| 5-HT | | 5-hydroxytryptamine (serotonin) |
| ABC | | ATP binding cassette |
| ACTH | | adrenocorticotropic hormone |
| ActR | | activin receptor |
| Amino acids | | |
| A | Ala | alanine |
| C | Cys | cysteine |
| D | Asp | aspartate |
| E | Glu | glutamate |
| F | Phe | phenylalanine |
| G | Gly | glycine |
| H | His | histidine |
| I | Ile | isoleucine |
| K | Lys | lysine |
| L | Leu | leucine |
| M | Met | methionine |
| N | Asn | asparagine |
| P | Pro | proline |
| Q | Gln | glutamine |
| R | Arg | arginine |
| S | Ser | serine |
| T | Thr | threonine |
| V | Val | valine |
| W | Trp | tryptophan |
| Y | Tyr | tyrosine |
| ANP | | atrial natriuretic peptide |
| apo | | apolipoprotein |
| ARA | | arachidonic acid |
| ATP | | adenosine-5'-triphosphate |
| βARK | | β-adrenergic receptor kinase |
| BNP | | brain natriuretic peptide |
| BoPCaR1 | | bovine parathyroid $Ca^{2+}$-sensing receptor |

| | |
|---|---|
| CAM | cell adhesion molecule |
| cAMP | cyclic AMP (adenosine-2':3'-phosphate or 2':3'-cyclic AMP) |
| CCK | cholecystokinin |
| cDNA | complementary DNA |
| CFTR | cystic fibrosis transmembrane conductance regulator |
| cGMP | cyclic GMP (guansoine-2':3'-phosphate or 2':3'-cyclic GMP) |
| CGRP | calcitonin gene-related peptide |
| CNP | C-type natriuretic peptide |
| CNS | central nervous system |
| CNTF | ciliary neurotrophic factor |
| CRE | cAMP-responsive element |
| CRF | corticotropin-releasing factor |
| cRNA | complementary RNA |
| C-terminus | carboxy-terminus |
| DAG | diacylglycerol |
| del | deletion |
| DNA | deoxyribonucleic acid |
| E-cadherin | epithelial-cadherin |
| EGF | epidermal growth factor |
| FGF | fibroblast growth factor |
| FHH | familial hypocalciuric hypercalcemia |
| FN | fibronectin |
| FMPP | familial male precocious puberty |
| FSH | follicle-stimulating hormone |
| GABA | γ-aminobutyric acid |
| GC | guanylate cyclase |
| G-CSF | granulocyte colony stimulating factor |
| GDB | Genome Data Base |
| GDP | guanosine-5'-diphosphate |
| GHRH | growth hormone-releasing hormone |

| | |
|---|---|
| GIP | gastric inhibitory polypeptide |
| GLP-1 | glucagon-like peptide-1 |
| GM-CSF | granulocyte-macrophage colony stimulating factor |
| GnRH | gonadotropin-releasing hormone |
| G protein | GTP-binding protein |
| GRE | glucocorticoid-responsive element |
| GRK | G protein-coupled receptor kinase |
| GRK1 | rhodopsin kinase |
| GRK2 | β-adrenergic receptor kinase (βARK)-1 |
| GRK3 | β-adrenergic receptor kinase (βARK)-2 |
| GRP | gastrin-releasing peptide |
| GTP | guanosine-5'-triphosphate |
| HCAM | homing-associated cell adhesion molecule |
| hCG | human choriogonadotropin |
| HCMV | human cytomegalovirus |
| HDL | high density lipoprotein |
| HlyB | hemolysin B |
| HVS | herpesvirus saimiri |
| ICAM | intercellular adhesion molecule |
| i.c.v. | intracerebroventricular |
| IDL | intermediate density lipoprotein |
| IFN | interferon |
| Ig | immunoglobulin |
| IGF | insulin-like growth factor |
| IL | interleukin |
| $IP_3$ | inositol 1,4,5-trisphosphate |
| $IP_4$ | inositol 1,3,4,5-tetrakisphosphate |
| $IP_6$ | phytic acid |
| kD | kilodaltons |
| KGF | keratinocyte growth factor |
| L-CAM | liver cell adhesion molecule |
| LDL | low density lipoprotein |

| | |
|---|---|
| LH | luteinizing hormone |
| LH/CG | luteinizing hormone/chorionic gonadotropin |
| LIF | leukemia inhibitory factor |
| LPA | lysophosphatidic acid |
| LRG | leucine-rich glycoprotein |
| MCP | monocyte chemoattractant protein |
| M-CSF | macrophage colony stimulating factor |
| MDR | multi-drug resistance |
| mGluR | metabotropic glutamate receptor |
| MGSA/GRO | melanoma growth stimulatory activity |
| MIP | macrophage inflammatory protein |
| *mrg* | *mas*-related gene |
| mRNA | messenger RNA |
| MSH | melanocyte-stimulating hormone |
| NAP | neutrophil activating protein |
| N-cadherin | neural cadherin |
| NCAM | neural-CAM |
| NDI | nephrogenic diabetes insipidus |
| NGF | nerve growth factor |
| NKA | neurokinin A (substance K) |
| NKB | neurokinin B (neuromedin K) |
| *N*-linked | Asn-linked |
| NPY | neuropeptide Y |
| NSHPT | neonatal severe hyperparathyroidism |
| N-terminus | amino-terminus |
| PACAP | pituitary adenylyl cyclase-activating peptide |
| PAF | platelet-activating factor |
| P-cadherin | placental cadherin |
| PCR | polymerase chain reaction |
| PDGF | platelet-derived growth factor |
| PECAM | platelet-endothelial cell adhesion molecule |
| PET | positron emission tomography |

| | |
|---|---|
| PG | prostaglandin |
| PI | phosphatidylinositol |
| PIP | phosphatidylinositol 4-monophosphate |
| $PIP_2$ | phosphatidylinositol 4,5-bisphosphate |
| $PIP_3$ | phosphatidylinositol 3,4,5-trisphosphate |
| PKA | protein kinase A (cAMP-dependent protein kinase) |
| PKC | protein kinase C |
| PLC | phospholipase C |
| PLD | phospholipase D |
| PTH/PTHrP | parathyroid hormone/parathyroid hormone related peptide |
| PYY | peptide YY |
| RANTES | regulated on activation, normal T expressed and secreted |
| RFLP | restriction fragment length polymorphism |
| RNA | ribonucleic acid |
| RP | retinitis pigmentosa |
| RTA | rat thoracic aorta |
| SCF | stem cell factor |
| SCLC | small cell lung cancer |
| sumatriptan | (3-[2-(dimethylamino)ethyl]-N-methyl-IH-indole-5-methane sulfonamide) (GR43175) |
| $T_3$ | triiodothyronine |
| $T_4$ | thyroxine |
| ter | termination codon |
| TGF | transforming growth factor |
| TM | transmembrane |
| TNF | tumor necrosis factor |
| TRH | thyrotropin-releasing hormone |
| TSH | thyroid-stimulating hormone |
| VCAM | vascular cell adhesion molecule |
| VIP | vasoactive intestinal peptide |
| ZP | *zona pellucida* |

# FOREWORD

In little over a decade the application of molecular biological techniques has revolutionized our understanding of the structural and functional diversity of receptors, complementing and enhancing physiological, pharmacological and biochemical investigations. There has been an explosion of interest and a wealth of information gained. It is not possible in one volume to acknowledge individually all publications which have contributed to our current knowledge of the largest single class of eukaryotic receptor, the G protein-coupled receptor. While we apologize to authors whose original contributions have not been specifically cited, we refer readers to relevant recent review articles mentioned in the text where many original research publications are cited and discussed.

# Acknowledgments

We thank colleagues for helpful discussions and critical commentary, in particular Professor Robert Graham, Dr. Peter Riek and Dr. Siiri Iismaa of the Victor Chang Cardiac Research Institute, Dr. Kieran Scott and Dr. Philip Conaghan of St. Vincent's Hospital, Sydney, and Dr. Peter Schofield and Dr. Majella Kelly of the Garvan Institute of Medical Research.

Much of the graphic art work in this publication was contributed by Ms. Melissa Coventry, whom we thank, along with Ms. Tricia Gleeson, for excellent secretarial assistance.

Tiina P. Iismaa, Ph.D.
Trevor J. Biden, Ph.D.
John Shine, Ph.D.
July, 1994

# CHAPTER 1

# CELL SURFACE RECEPTORS AND THE G PROTEIN-COUPLED RECEPTOR SUPERFAMILY

Transduction of extracellular signals across the plasma membrane to the intracellular environment is achieved by the interaction of regulatory molecules with specific membrane-spanning cell surface receptors. Interaction of an appropriate activating ligand with the receptor at the external face of the cell results in the generation of an intracellular signal. Cell surface receptors are able to distinguish their specific ligands from the multitude of other bioactive factors in the extracellular milieu. In response to specific activation, such receptors function in the transmission, amplification and integration of extracellular signals through a variety of intracellular mechanisms to control cellular functioning.

A wide range of neurotransmitters, neuropeptides, polypeptide hormones, inflammatory mediators and other bioactive molecules transduce their signal to the intracellular environment by specific interaction with a class of receptor that relies upon interaction with intracellular guanosine-5'-triphosphate (GTP)-binding proteins (G proteins) for activation of intracellular effector systems. In recent years, molecular cloning approaches have allowed the identification of several hundred discrete G protein-coupled receptor molecules, and it has been established that approximately 80% of known hormones and neurotransmitters activate cellular signal transduction mechanisms by activating G protein-coupled receptors.[1] This renders the G protein-coupled receptor superfamily, with its common structural and functional features, the largest single class of eukaryotic receptor.

In this chapter we review, within the context of cell surface receptor superfamiles which have been described to date, the molecular architecture and structural diversity of the G protein-coupled receptor superfamily.

## CELL SURFACE RECEPTORS

Based on structural and functional criteria, three broad categories of cell surface receptor which recognize specific molecules may be defined. These comprise receptors involved in cellular adhesion processes, receptors which capture and convey ligands to appropriate intracellular destinations and receptors which initiate a sequence of intracellular transducing signals when activated by their ligand.

## 1. CELL ADHESION MOLECULES

At least five classes of receptor are known to be involved in cell-extracellular matrix interactions, intercellular adhesion and cell-adhesion-associated intercellular communication. These are the cell adhesion molecules (CAMs), the calcium dependent adhesion molecules (cadherins), the selectins, the integrins and the homing-associated cell adhesion molecule (HCAM). The structural features of representative members of these receptor classes are shown schematically in Figure 1.1a.

### i. CAMs

The CAMs are members of the immunoglobulin gene superfamily that also gives rise to soluble antigen recognition molecules and antigen-specific and antigen nonspecific cell surface receptors.[2-4] The CAMs are transmembrane glycoproteins which function primarily in cell-cell adhesion and are characterized by the presence of one or more immunoglobulin-like domains in the extracellular region. They mediate calcium-independent binding through either homophilic or heterophilic interactions and include neural-CAM (NCAM) which occurs on neurons and glia,[5] the intercellular adhesion molecules ICAM-1, expressed by a wide variety of cells in response to inflammatory mediators and ICAM-2, expressed constitutively by endothelial cells, the platelet-endothelial cell adhesion molecule PECAM-1, the vascular cell adhesion molecule VCAM-1 and the lymphocyte function-associated antigens LFA-2 and LFA-3. Among the CAMs, ICAM-1 serves also as the receptor for rhinovirus.[4,6]

### ii. Cadherins

The cadherins are transmembrane glycoproteins which mediate vertebrate cell adhesion primarily through homophilic binding in a $Ca^{2+}$-dependent manner. A number of subclasses of cadherins and related

molecules have been described, with four subclasses of cadherins having been characterized extensively at the molecular level. These are E-cadherin (epithelial-cadherin or uvomorulin), N-cadherin (neural cadherin), P-cadherin (placental cadherin) and L-CAM (liver cell adhesion molecule). Each subclass exhibits a unique tissue distribution. In addition, many cell types coexpress multiple cadherin subclasses in varying combinations, allowing delineation of cell-specific expression profiles for members of this receptor family.[7-12]

### iii. Selectins

The selectins are cell surface carbohydrate-binding proteins, or lectins, and comprise a family of three members. L-selectin is expressed constitutively on lymphocytes and plays a central role in lymphocyte recirculation through peripheral lymph nodes. Both E-selectin, which is expressed on endothelial cells in response to inflammatory stimuli and P-selectin, which is stored in granules and transported to the cell surface of endothelial cells and platelets in response to injury or inflammatory mediators, are involved in the migration of leukocytes in developing inflammatory reactions. These receptors mediate the adhesion of leukocytes to endothelium and platelets during inflammation and clotting.[13-17]

### iv. Integrins

The integrins are a large family of transmembrane glycoproteins with widespread tissue distribution which function in cell adhesion and in the transduction of molecular signals from the extracellular matrix.[18-21] They are heterodimeric, comprising an $\alpha$ and $\beta$ subunit, of which the $\alpha$ subunit is the primary determinant of ligand specificity, and will interact with ligands containing the tripeptide sequence Arg-Gly-Asp (RGD). To date, 12 $\alpha$ and 8 $\beta$ molecular variants have been described. Several of these variants are able to generate heterodimers with more than one complementary chain and at least 20 different integrin heterodimers have been characterized, representing a diversity of ligand specificity and characteristic cell or tissue distribution.[22]

## v. H-CAM

H-CAM (CD44) is a glycoprotein lectin of widespread tissue distribution which exists as several isoforms. It functions as a receptor for hyaluronate and in this capacity is involved in the anchoring of cells to extracellular matrix and basement membrane and in the regulation of cell motility and shape. It is essential for the regulation of normal immune cell function and plays a role in lymphocyte recirculation between blood and lymphoid organs and in the initiation and perpetuation of inflammatory responses. It has also been implicated in tumor cell invasiveness.[6,23-27]

## 2. UPTAKE RECEPTORS AND TRANSPORTERS

A wide variety of cell surface receptors internalize their ligand to effect the uptake of metabolites or diffusible molecules into the cell. Included in this structurally and functionally diverse category of receptors are the low density lipoprotein (LDL) and transferrin receptors, the glucose, amino acid, urea and plasma membrane neurotransmitter transporters, and members of the adenosine nucleotide (ATP) binding cassette (ABC) superfamily of transporter proteins. Representative receptors of this functional class of cell surface molecule are depicted schematically in Figure 1.1b.

## i. Low density lipoprotein (LDL) receptor

Cellular uptake of cholesterol is achieved by lipoprotein receptor-mediated endocytosis of cholesteryl ester-carrying plasma lipoproteins. The prototypic lipoprotein receptor is the LDL receptor (Fig. 1.1b), a transmembrane glycoprotein that binds apolipoprotein (apo) B-100, the sole protein constituent of LDL, and apo E, the protein component of intermediate density lipoprotein (IDL) and a subclass of high density lipoprotein (HDL). Receptors are localized on the cell surface in clathrin-

coated pits. Ligand binding is followed by internalization of the ligand-receptor complex into coated vesicles. Dissociation of ligand from receptor occurs in acidified endosomes; the LDL is delivered to lysosomes for degradation and receptors are recycled to the cell surface.[28-30]

## ii. Transferrin receptor

The transferrin receptor (Fig. 1.1b) is an integral membrane protein composed of two identical, disulfide-linked subunits.[31] It controls the supply of iron to the cell through the binding of transferrin, the major iron-carrier protein in serum. Iron-bearing transferrin is internalized by receptor-mediated endocytosis, iron is dissociated within acidic clathrin-coated endosomes and both apotransferrin and receptor are recycled to the cell surface.[32,33]

## iii. Glucose transporters

Two classes of carrier protein for D-glucose occur in mammalian cells (Fig. 1.1b). The facilitative glucose transporters, denoted GLUT1, 2, 3, 4 and 5, accelerate the transport of glucose down its concentration gradient by facilitative diffusion. They exhibit distinct tissue distributions which reflect specific functional properties. These transporters characteristically contain two segments comprising six putative membrane-spanning domains which are separated by a large hydrophilic cytoplasmic region. Both amino (N)- and carboxy (C)-terminal segments of the transporters are predicted to be cytoplasmic.[34-36]

The $Na^+$/glucose cotransporter, by coupling the uptake of glucose with the uptake of $Na^+$ ions, transports glucose against its concentration gradient. It is expressed on absorptive epithelial cells of the small intestine and functions in the uptake of dietary glucose. This transporter is a glycoprotein predicted to traverse the plasma membrane 11 times, with N- and C-termini located on the cytoplasmic and extracellular faces of the membrane, respectively. It exhibits sequence features shared by a number of $Na^+$-coupled transporters, including mammalian neutral amino acid

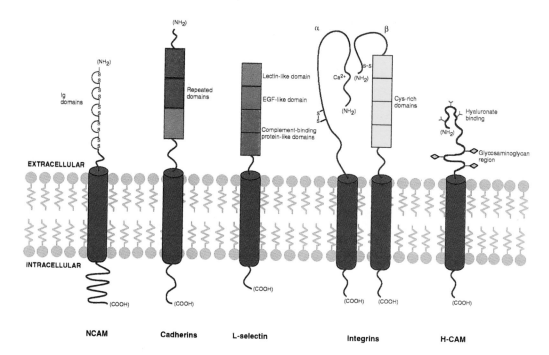

*Fig. 1.1a. Representative members or generalized structure of each of the five classes of cell adhesion molecule. The structure of CAMs is represented by neural-CAM (NCAM), the extracellular domain of which contains five internally homologous segments, connected by intrachain disulfide bonds (−SS−), that fold as immunoglobulin (Ig) domains. The generalized structure of cadherins depicts the three characteristic repeated domains in the extracellular region, that carry sequences involved in $Ca^{2+}$ binding. The selectin family of cell adhesion molecules is represented by L-selectin and shown schematically are EGF-like domains and lectin-like domains found in the extracellular segment of all three selectins. Within the extracellular region, L-selectin contains two complement binding protein-like domains, in contrast to the six and nine such domains found in the extracellular segment of E-selectin and P-selectin, respectively. The integrins comprise an α and a β subunit, each of which exhibits intrachain disulfide bonding (−S-S−). Regions thought to bind $Ca^{2+}$ within the extracellular domain of the α subunit are shown, as are the four repeats of a 40 amino acid Cys-rich sequence motif that occur in the β subunit. H-CAM molecules are characterized by potential glycanation sites for chondroitin sulfate or heparan sulfate (◊) within the glycosaminoglycan region of the extracellular domain and sites for N-linked-glycosylation (Y) within the hyaluronate binding domain. Figures are not drawn to scale.*

*Fig 1.1b. (opposite) Representative members or generalized structure of uptake receptors and transporters. The extracellular domain of the low density lipoprotein (LDL) receptor comprises three domains, with negatively charged groups within the N-terminal ligand binding domain attributable to Cys-rich repeat sequences, the epidermal growth factor (EGF) precursor homology domain containing three growth factor-like repeats and O-linked carbohydrate groups (−) attached in the membrane-proximal region. The transferrin receptor comprises two identical subunits linked by disulfide bonding (−S-S−). Sites of attachment of high mannose oligosaccharides and complex oligosaccharides within the C-terminal extracellular domain are shown. The N-terminal intracellular segment contains sites for attachment of fatty acid (FA) moieties and phosphorylation sites (P). Facilitative glucose, $Na^+$-dependent glucose and urea transporters span the plasma membrane 12, 11 and 10 times, respectively, and contain potential sites for N-linked glycosylation (Y) on the first, third and fifth extracellular loops, respectively. Representative of the ATP-binding cassette (ABC) transporter family is the cystic fibrosis transmembrane conductance regulator (CFTR), which spans the plasma membrane 12 times and contains NBFR1, NBFR2 and R cytoplasmic domains.*

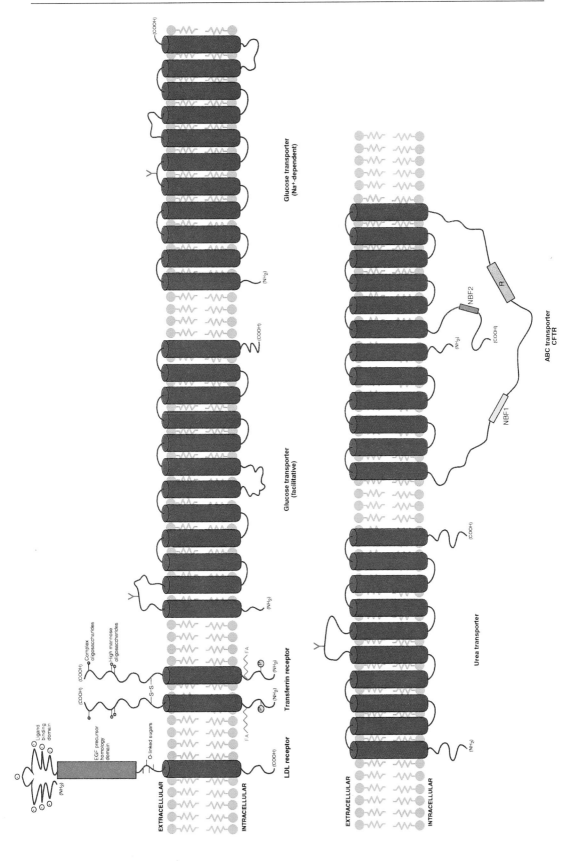

transporters (see below), nucleoside, *myo-*inositol and phosphate cotransporters and bacterial cotransporters for proline, pantothenate and glutamate.[37]

### iv. Amino acid transporters

The uptake of L-amino acids into the cell is accomplished by functionally and biochemically distinct amino acid transporters, encompassing both $Na^+$-dependent and $Na^+$-independent systems.[38] In recent years, the molecular architecture of some of these transporter proteins has begun to be elucidated.

The mammalian $Na^+$-dependent neutral amino acid transporter (system A)[37] bears significant sequence homology with other $Na^+$-dependent transporter proteins, including the $Na^+$/glucose cotransporter,

and is presumed to exhibit similar topological organization within the plasma membrane (see above; Fig. 1.1b).

The principal transporter of the cationic amino acids arginine, lysine and ornithine across the plasma membrane is the $Na^+$-independent $y^+$ transporter. It is an integral membrane protein with 14 putative transmembrane domains. The murine $y^+$ transporter has been shown also to have been subverted for use as the receptor for ecotropic host-range murine leukemia viruses.[39-41]

The $Na^+$-independent uptake of cystine and dibasic and neutral amino acids is mediated by the broad spectrum $b_{o,+}$ transporter. Molecular cloning has been used to isolate two closely related transporter molecules, each of which has four putative transmembrane domains and exhibits functional

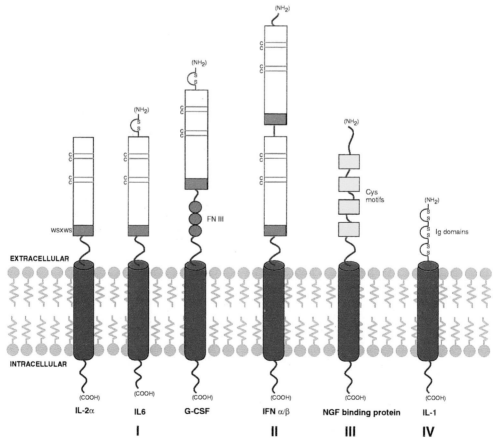

Fig 1.1c. Representative members of cytokine receptor types I, II, III and IV. A repeated (Trp-Ser-X-Trp-Ser) motif characteristic of type I receptors is shown (WSXWS), as are Cys residues (C) conserved in type I and type II receptors. Immunoglobulin-like domains (Ig), fibronectin III (FN III) domains, and the Cys motifs that characterize type III cytokine receptors, are identified.

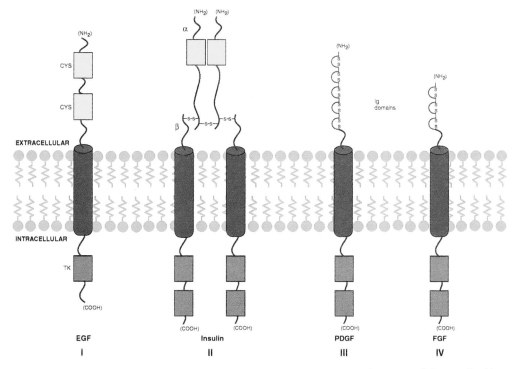

Fig. 1.1d. Representative members of tyrosine kinase receptor types I, II, III and IV. Intracellular tyrosine kinase domains (TK) found in all tyrosine kinase receptors, Cys-rich repeat regions (CYS) occurring in type I and II receptors, and immunoglobulin-like (Ig) domains occurring in type III and IV receptors, are shown.

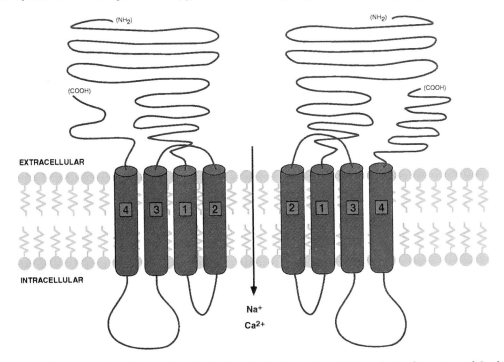

Fig. 1.1e. Representation of two subunits of a ligand-gated ion channel. The relative disposition of the four transmembrane domains within each subunit, with the second consecutive transmembrane domain lining the ion channel, is depicted schematically. The complete ion channel comprises five subunits. The arrow denotes flow of ions.

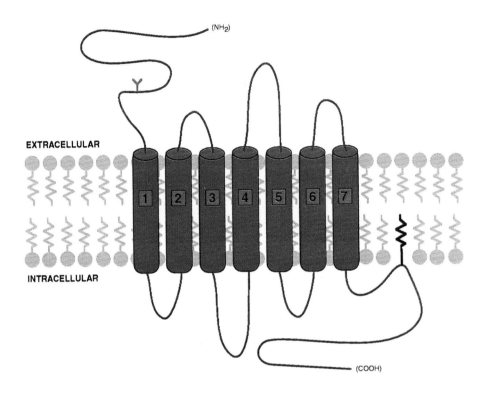

Fig. 1.1f. Schematic representation of a G protein-coupled receptor. The single polypeptide chain traverses the plasma membrane seven times and transmembrane domains are numbered consecutively. One or more sites for N-linked glycosylation (Y) may occur in the extracellular N-terminal domain or in extracellular loops. The C-terminal cytoplasmic domain may be attached to the internal face of the plasma membrane through palmitoylation.

features of this transport system. However, it is not clear whether these transporter molecules are solely responsible for amino acid uptake mediated by this system, or represent only part of a functional transporter complex.[42-44]

### v. Urea transporter

The urea transporter (Fig. 1.1b) plays a central role in fluid balance and nitrogen metabolism by specifically transporting the waste metabolite urea across cell membranes within kidney medullary collecting ducts. The transporter protein comprises ten putative membrane-spanning domains with N- and C-terminal domains located intracellularly. It bears no sequence homology to any other known transporter protein.[45]

### vi. Neurotransmitter transporters

Plasma membrane neurotransmitter transporters function in the re-uptake and recycling of neurotransmitters at the synapse and comprise two structurally and mechanistically distinct classes. The first of these are the plasma membrane carriers for γ-aminobutyric acid (GABA), noradrenaline/adrenaline, dopamine, serotonin, choline, proline, glycine, betaine and taurine, which couple neurotransmitter uptake to the influx of $Na^+$ and $Cl^-$ ions. Transporter proteins of this class are integral membrane glycoproteins comprising a single polypeptide chain which traverses the plasma membrane 12 times. N- and C-termini are cytoplasmic.

To the second class of plasma membrane transporter belong the carriers for the excitatory amino acids glutamate and aspartate, where neurotransmitter uptake is coupled to the co-transport of $Na^+$ and the counter-transport of $K^+$, with no apparent dependence on $Cl^-$ ions. These transporter proteins are also integral membrane glycoproteins but structural predictions have yielded conflicting results, with membrane topologies incorporating 6, 8 and 10 transmembrane domains having been reported. The two classes of plasma membrane neurotransmitter transporter are functionally and pharmacologically distinct from the proton-dependent vesicular carriers that occur in the membranes of synaptic vesicles.[46-48]

## vii. ATP binding cassette (ABC) transporter superfamily

The ATP binding cassette (ABC) transporters occur in both prokaryotic and eukaryotic systems and are involved in the translocation of substrates as diverse as ions, heavy metals, amino acids, peptides, proteins, sugars and anti-cancer drugs across biological membranes. Members of the superfamily include bacterial periplasmic permeases which mediate nutrient uptake, the *Escherichia coli* inner membrane protein hemolysin B (HlyB) which functions in secretion of the enterotoxin hemolysin A, and eukaryotic proteins such as the P-glycoprotein efflux pump associated with multi-drug resistance (MDR) in tumors, the *Plasmodium falciparum* pfMDR 1 gene product that is implicated in chloroquine resistance of the malarial parasite, the major histocompatibility complex-associated transporter of antigenic peptides and the cystic fibrosis transmembrane conductance regulator (CFTR) protein (Fig. 1.16).

There is extensive sequence homology and a highly conserved domain organization among members of the superfamily. The functional transporter may be composed of several subunits, each representing one or two of the functional domains, or may comprise a single polypeptide chain. The predicted protein structure has two hydrophobic domains, each of which has five or six putative α-helices which span the membrane, and are believed to bind the substrate and mediate its passage across the membrane. There are two domains, denoted NFB-1 and -2 in the CFTR, which couple ATP hydrolysis to transport and are associated with the cytoplasmic face of the membrane. In the case of the CFTR, there is a fifth domain, the R subunit, which is also considered to be cytoplasmic and to serve a regulatory role.[49-51]

## 3. SIGNAL TRANSDUCTION SYSTEMS

Cell surface receptors which are components of signal transduction systems primarily interact with soluble extracellular ligands. Receptor activation results in modulation of intracellular functioning. This category of receptors comprises molecules of considerable structural and functional diversity and includes the cytokine receptors, the tyrosine kinase and serine/threonine kinase, protein tyrosine phosphatase and guanylate cyclase receptor families, the ligand-gated ion channels and the G protein-coupled receptors.

### i. Cytokine receptors

Cytokine receptors are transmembrane glycoproteins which may be classified into several distinct families on the basis of amino acid sequence features of the extracellular domain or intrinsic catalytic activity in the cytoplasmic domain of the receptor.[52-57] Representative members of each family are shown schematically in Figure 1.1c.

Type I cytokine receptors include the hematopoietic growth factor receptors such as those for interleukin (IL)-2α and β, IL-3, IL-4, IL-5, IL-6, IL-7, IL-9, granulocyte colony stimulating factor (G-CSF), granulocyte-macrophage colony stimulating factor (GM-CSF), erythropoietin, leukemia inhibitory factor (LIF) and oncostatin M, as well as receptors for ciliary neurotrophic factor (CNTF), growth hormone and prolactin. They may be monomeric, homodimeric, heterodimeric, multimeric, may associate with other membrane proteins or be differentially glycosylated to effect high affinity ligand binding and signal transduction.[54,58-60]

These receptors share structural homology within a stretch of approximately 200 amino acid residues of the extracellular ligand binding domain, with conservation of Cys residues and a repeated Trp-Ser motif (Trp-Ser-X-Trp-Ser). The extracellular domain includes immunoglobulin-like domains in IL-6 and G-CSF receptors, and fibronectin III domains in G-CSF, LIF and oncostatin M receptors.

Type II cytokine receptors include the receptor for IL-10, as well as receptors that mediate the antiviral activity of the interferons (IFN)-α and -β, which interact with a common receptor, and IFN-γ. While these receptors appear to be structurally and evolutionarily related to the type I cytokine receptors, they do not exhibit conservation of the (Trp-Ser-X-Trp-Ser) motif characteristic of type I receptors.[60]

Type III cytokine receptors are characterized by Cys-rich extracellular domains and include the receptors for tumor necrosis factor (TNF)-α and -β and the low affinity nerve growth factor (NGF) receptor (NGF binding protein). Other members of the family include the B-cell antigen CD40, the T-cell antigens CD27 and OX40, and other cell surface proteins including the apoptosis inducer antigen FAS, the induced T-cell-derived cDNA 4-1BB and CD30, a marker for Hodgkin's lymphoma.[56,60]

Type IV cytokine receptors are members of the immunoglobulin superfamily[2-4] and are typified by the IL-1 receptors. Immunoglobulin-like domains occur also in the extracellular domain of receptors for the cytokines macrophage colony stimulating factor (M-CSF) and stem cell factor (SCF), and the growth factors platelet-derived growth factor (PDGF) and fibroblast growth factor (FGF) α and β. However, these receptors also exhibit intrinsic tyrosine kinase activity and will be discussed below in the context of the tyrosine kinase receptor family.

Receptors for the cytokines IL-8 and transforming growth factor (TGF)-β represent two additional types of cytokine receptor which, in a structural and functional context, belong to the G protein-coupled receptor and serine/threonine kinase families, respectively.

The features of these receptors will be discussed below in the relevant sections.

## ii. Tyrosine kinase receptors

Members of this receptor family have a large extracellular ligand binding domain, a single transmembrane domain and are characterized by the presence of an intrinsic intracellular enzymatic domain which has tyrosine kinase activity and catalyses the phosphorylation of tyrosine residues in response to ligand binding. Tyrosine kinase receptors may be classified into subclasses on the basis of sequence similarity and distinguishing structural features.[61-63] Schematic representations are given in Figure 1.1d.

Subclass I receptors are monomeric, have two Cys-rich repeat regions in the extracellular domain and include the epidermal growth factor (EGF) receptor and the HER2/*neu* receptor-like gene product.

Subclass II receptors function as heterotetrameric disulfide-linked structures comprising two extracellular α subunits which have a single Cys-rich region and contribute to the ligand binding domain and two β subunits that span the plasma membrane and carry the intracellular tyrosine kinase domain. Belonging to this class of receptor are the insulin and insulin-like growth factor (IGF)-I receptors.

Subclass III receptors are monomeric and include receptors for PDGF, CSF-1 and the cytokines M-CSF and SCF. These receptors lack the Cys-rich repeat clusters seen in the extracellular domain of subclass I and II tyrosine kinase receptors but are characterized by the presence of five Cys-containing immunoglobulin-like domains in the extracellular domain and interruption of the intracellular catalytic domain by an insertion sequence of hydrophilic amino acid residues.

Subclass IV receptors exhibit two or three immunoglobulin-like domains in the extracellular domain and include the FGF and keratinocyte growth factor (KGF) receptors.[64,65]

## iii. Serine/threonine kinase receptors

The TGF-β gene family comprises a family of peptide mediators which includes isoforms of TGF-β, activins, inhibins and bone morphogenic factor. Type II and type V TGF-β receptors and activin receptors (ActR)-II and -IIB represent members of a family of single-pass transmembrane receptors which contain an intracellular serine/threonine kinase catalytic domain.[66-70]

A putative receptor belonging to this receptor family[71] which also contains serine/threonine kinase domains, has recently been suggested to represent the type I TGF-β receptor.[72]

## iv. Protein tyrosine phosphatase receptors

Protein tyrosine phosphatases are the catalytic counterpart to protein tyrosine kinases. The family of transmembrane protein tyrosine phosphatases comprises at least 10 putative members, each exhibiting an N-terminal extracellular domain, a single transmembrane region and one or two protein tyrosine phosphatase domains located within the cytoplasmic C-terminal segment of the protein. The most intensively studied of these receptors is CD45, a structurally heterogeneous family of hematopoietic cell-specific surface glycoproteins, which plays a crucial role in signaling events in T lymphocytes, B lymphocytes and natural killer cells.[69,73-77]

## v. Guanylate cyclase receptors

Guanylate cyclase (GC) receptors comprise an extracellular ligand binding domain and intracellular protein kinase-like and cyclase catalytic domains separated by a single putative transmembrane domain. Receptor activation results in the generation of cyclic GMP (cGMP) and activation of an intracellular cGMP cascade. Members of this receptor family include the natriuretic peptide receptors GC-A, which exhibits high affinity for atrial natriuretic peptide (ANP) and brain natriuretic peptide (BNP), and GC-B, which is selective for C-type natriuretic peptide (CNP), as well as GC-C, which binds heat stable enterotoxins/guanylin and RetGC, suggested to function as the photoreceptor guanylate cyclase.[78-80]

The ANP clearance receptor, which binds all three natriuretic peptides, is also a member of this family. Originally named for its role in clearance of natriuretic peptides from plasma, it contains neither a protein kinase-like nor a cyclase catalytic domain, but has been suggested to mediate some of the biological actions of these peptides.[81]

## vi. Ligand-gated ion channels

Ligand-gated ion channels mediate the rapid action of neurotransmitters at the synapse, with the selective passage of ions through the channel into the cell occurring within microseconds of activation of the receptor by ligand binding. Propagation or inhibition of neuronal impulses is achieved through the selective permeability of ligand-gated ion channel receptors to either cations or anions, respectively. Receptors for the excitatory neurotransmitters acetylcholine, glutamate and serotonin allow the passage of cations, generating an ion current that results in depolarization of the postsynaptic cell. The inhibitory neurotransmitters GABA and glycine-activated receptors which cause membrane hyperpolarization, due to the inward movement of chloride anions. The response is rapid and transient, with a latency for channel opening of microseconds. Receptors may become desensitized and refractory to further stimulation if agonist remains within the synaptic cleft for more than a few milliseconds. Responsiveness is re-acquired within a time frame of seconds to minutes.[82-86]

The most intensively studied ligand-gated ion channel is the nicotinic acetylcholine receptor, which has been known for some time to occur as a heteropentameric structure comprising $\alpha_2\beta\gamma\delta$ subunits.[87] The more recent molecular description of component subunits of other ligand-gated ion channels has clearly identified the existence of a superfamily which broadly represents this functional class of receptor. Individual channel subunits comprise a single polypeptide chain which is predicted to have an hydrophobic signal sequence, a long extracellular N-terminal domain containing multiple glycosylation sites, four or possibly five membrane-spanning domains,[86] and a C-terminal segment which is relatively short in most cases (Fig. 1.1e). In addition, a further level of complexity has begun to be appreciated with the identification of molecular variants of several of the receptor subunits. It is becoming increasingly clear that receptor subunit multiplicity provides the molecular basis for the distinguishing electrophysiological and pharmacological features of native receptor subtypes. For example, the GABA$_A$ receptor antagonist bicuculline blocks virtually all GABA$_A$ receptors, preventing inhibitory feedback and leading to seizures, thereby rendering it of little therapeutic value. Benzodiazepines such as valium, on the other hand, are clinically useful for enhancement of the effects of GABA since they bind only GABA$_A$ receptors which contain the $\gamma_2$ subunit and lack the $\alpha_6$ subunit.[88,89]

On the basis of amino acid sequence features, members of the ligand-gated ion channel superfamily may be classified into two families. This classification does not correlate with the functional classification of receptors based on their selective permeability to cations or anions. Subunits of the excitatory nicotinic acetylcholine and serotonin receptors, and the inhibitory GABA$_A$ and glycine receptors, share considerable sequence homology, particularly in the membrane-spanning domains. The homology between subunits of different channels is approximately 20-40%, while between subunits of individual channels is usually greater than 40%. Glutamate receptor subunits share no overall primary sequence homology with the other ligand-gated ion channel subunits, but do between themselves and therefore constitute a discrete family.[90]

## vii. G protein-coupled receptors

G protein-coupled receptors are integral membrane proteins which occur in a wide range of organisms and interact with a diversity of ligands. The receptor superfamily includes rhodopsin and the visual color pigments, in which photons of light induce isomerization of the covalently-bound chromophore, retinal, as well as receptors which bind neurotransmitters, neuropeptides, polypeptide hormones, inflammatory mediators and other bioactive molecules to initiate intracellular signaling events. All members of the G protein-coupled receptor superfamily described to date comprise a single polypeptide chain which characteristically contains seven stretches of mostly hydrophobic residues of 20 to 30 amino acids, linked by hydrophilic domains of varying length. They mediate signal transduction by interaction with ligand at the extracellular side of the membrane, followed by activation of intermediary receptor-associated intracellular G proteins, which in turn act upon effector molecules.

It should be noted at this point that not all receptors that activate G proteins are members of the G protein-coupled receptor superfamily. The activation of heterotrimeric G proteins has been implicated in signal transduction mediated by several tyrosine kinase receptors (see Section 3(ii), above), including the receptors for EGF, insulin, insulin-like growth factor II (IGF-II) and CSF-1.[91-94] For example, the receptor for IGF-II has been shown to directly activate the G protein G$_{i-2}$, with subsequent activation of a calcium-permeable channel. A 14 amino acid sequence within the cytoplasmic C-terminal portion of this receptor has a similar charge distribution to domains in adrenergic receptors that are implicated in G protein-coupling.[95,96] Such structural features which may be important in receptor-G protein interactions will be discussed in detail in chapter 3.

A common topographical organization (Fig. 1.1f) and three-dimensional structure (Fig. 1.2) is assumed for the integral membrane regions of all members of the G protein-coupled receptor superfamily. Seven $\alpha$-helical segments of mostly hydrophobic amino acid residues, connected by alternating cytoplasmic and extracellular hydrophilic loops, are considered to form transmembrane (TM) domains which occur as $\alpha$-helical "cylinders" traversing the membrane and are arranged such that a central pore is formed on the extracellular surface. This structure was originally proposed on the basis of the analogous hydropathic features exhibited by bovine rhodopsin and the $\beta_2$-adrenergic receptor, which are G protein-coupled receptors, and bacteriorhodopsin, the purple membrane photoprotein of the prokaryote *Halobacterium halobium*,[97-102] and is consistent with general features that can be derived for all members of the superfamily.[103,104] It has subsequently been demonstrated that in both rhodopsin and the $\beta_2$-adrenergic receptor, the N-terminus occurs on the extracellular side of the plasma membrane and the C-terminus is intracellular.[105,106]

A variety of computer modeling approaches have been used in efforts to delineate the relative disposition of each of the TM helices and to construct three-dimensional models of these receptors.[107-114]

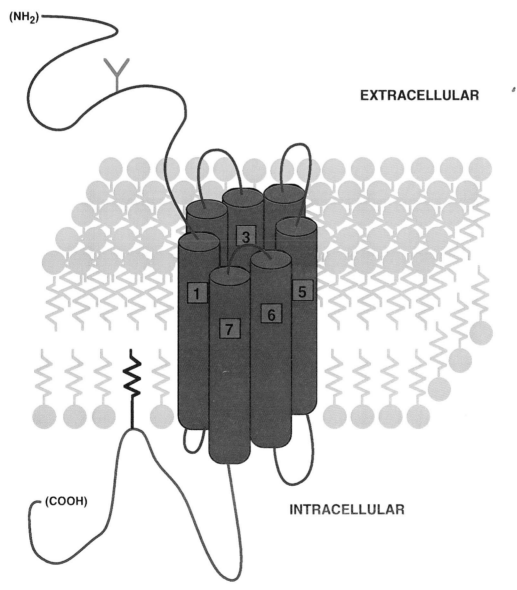

Fig. 1.2. Schematic representation of the three-dimensional structure of a G protein-coupled receptor. Putative transmembrane domains are depicted as cylinders and their relative disposition within the membrane to form a central pore on the extracellular surface is shown. One or more potential sites for N-linked glycosylation (**Y**) may occur in the extracellular domain and attachment of the C-terminal cytoplasmic domain of the receptor to the internal face of the plasma membrane through palmitoylation is depicted schematically.

Most of these models have been generated using bacteriorhodopsin as a template. A proposed structure consistent with the projection map of rhodopsin obtained by electron crystallography of two-dimensional crystals[115,116] suggests that the structure of G protein-coupled receptors differs from that of bacteriorhodopsin but that the general three-dimensional arrangement of TM helices is similar.[117] Clearly, the derivation of a definitive structure for the G protein-coupled receptor superfamily awaits the attainment and analysis of three-dimensional crystallographic data.

G protein-coupled receptors rely upon interaction with a G protein complex in the plasma membrane to achieve regulation of intracellular effector systems (Fig. 1.3). Agonist binding induces a conformational change in the receptor which is transmitted to the G protein. This results in activation of the G protein, which in turn activates a separate effector such as an ion channel or an enzyme. The involvement of a G protein in the transmission of signal from receptor to effector renders this mode of signal transduction slower than that achieved by ligand-gated ion channels, where both receptor and effector occur in the one oligomeric transmembrane protein.

G proteins are heterotrimeric molecules, comprising a variable $\alpha$ subunit and relatively invariant $\beta$ and $\gamma$ subunits. On activation, guanosine-5'-diphosphate (GDP), which is normally bound to the $\alpha$ subunit, is released and replaced by GTP. The $\alpha$ subunit, with bound GTP, separates from the $\beta\gamma$ subunits and this generates a bifurcating signal within the cell, since both the $\alpha$- and $\beta\gamma$-subunits are able to bind to effector systems in the membrane, such as ion channels or enzymes, to either stimulate or inhibit their activity.[1,118] The specificity of signal transduction is determined largely by the specificity of G protein coupling, since different G protein subunits preferentially stimulate particular effectors. Subsequently, the G protein $\alpha$ subunit, which contains an integral GTPase activity, hydrolyses GTP, reverts to the GDP-bound, inactive state, and dissociates from the ion channel or enzyme. Signal amplification occurs as a consequence of the ability of a single receptor to activate many G protein molecules and from the initiation of several subsequent catalytic cycles of effector enzymes (see chapter 2).[1,118]

G protein-coupled receptors which transduce their signal by coupling directly to ion channels through G proteins elicit an alteration in membrane potential with a latency to onset of at least 30 milliseconds. This is not as rapid as that achieved by activation of ligand-gated ion channels, but generally has an overall longer duration, on the order of hundreds of milliseconds. The G protein-coupled receptor-mediated activation of enzymes such as adenylyl cyclase, guanylate cyclase, phospholipases C and $A_2$ and phosphodiesterases results in the generation of intracellular second messenger molecules such as cyclic AMP (cAMP), cGMP, diacylglycerol (DAG), inositol 1,4,5-trisphosphate ($IP_3$), and arachidonic acid and associated metabolites. This sets in motion a cascade of events which continues through the capability of second messenger molecules to elicit a variety of subsequent effects. These may include alteration of the activity of various protein kinases or the mobilization of calcium from intracellular stores, which may in turn affect the activity of ion channels or other enzymes. Effects such as the phosphorylation of target proteins occur within a time frame of seconds to minutes after receptor activation. For example, the cAMP-mediated increase of $Ca^{2+}$ current in the heart relies upon phosphorylation of $Ca^{2+}$ channels. In contrast to direct G protein coupling to ion channel activity, this indirect ion channel activation has a latency to onset of several seconds and $Ca^{2+}$ current is observed to increase over 30 seconds. The slower onset and increasing intensity of response reflect the necessity for the generation of the second messenger cAMP, followed by activation of increasing numbers of $Ca^{2+}$ channels by phosphorylation by the activated protein kinase. Longer term responses to receptor activation may include modulation of transcriptional

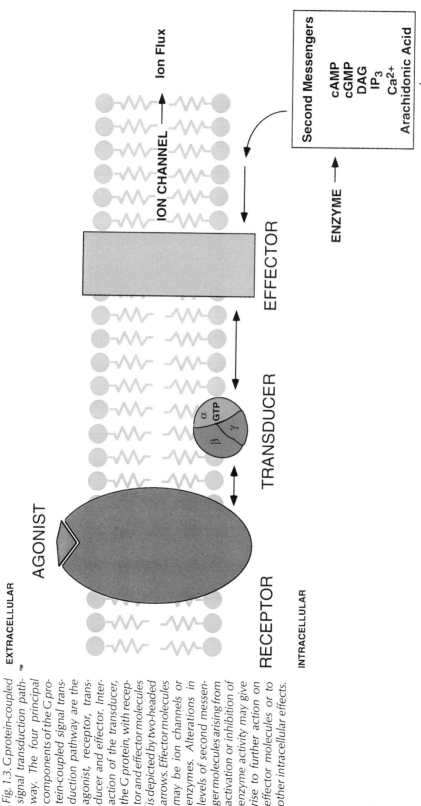

Fig. 1.3. G protein-coupled signal transduction pathway. The four principal components of the G protein-coupled signal transduction pathway are the agonist, receptor, transducer and effector. Interaction of the transducer, the G protein, with receptor and effector molecules is depicted by two-headed arrows. Effector molecules may be ion channels or enzymes. Alterations in levels of second messenger molecules arising from activation or inhibition of enzyme activity may give rise to further action on effector molecules or to other intracellular effects.

activity in the cell nucleus, resulting in an altered phenotype several hours after receptor activation.[82,83,119]

The involvement of G proteins in the signal transduction process and the diversity of effectors with which they interact implies that activation of a particular receptor at the cell surface may result in a number of subsequent intracellular events. An example of such divergence of ligand action is seen with the action of noradrenaline on $\alpha_2$-adrenergic receptors in guinea pig submucous neurons, resulting in both opening of $K^+$ channels and closure of $Ca^{2+}$ channels.[83] On the other hand, convergence of response to receptor activation may occur if a number of different receptors expressed on a particular cell couple to the same effector molecule. This is seen with the convergence of activation of muscarinic acetylcholine, adrenergic, adenosine, dopamine, GABA, serotonin, galanin, opioid and somatostatin receptors onto the same inward rectifying neuronal $K^+$ channel, or the inhibition of N-type $Ca^{2+}$ channels by both pertussis toxin-sensitive and -insensitive G proteins coupled to a variety of receptors. In such instances the relative effectiveness of a particular ligand in eliciting a response may be determined in part by selectivity in coupling to different classes of G protein.[83,119]

An additional level of complexity in receptor-mediated regulation of cellular functioning is seen among the neurotransmitters acetylcholine, ATP, glutamate, GABA and serotonin, which interact with receptors belonging to both ligand-gated ion channel and G protein-coupled receptor superfamilies. This provides the means for both rapid effects on membrane potential and longer-term modulation of ion channel or metabolic activity to be achieved.

# STRUCTURAL AND FUNCTIONAL FEATURES OF G PROTEIN-COUPLED RECEPTORS

While G protein-coupled receptors exhibit a characteristic hydropathy profile, with a relatively constant length of primary sequence comprising the TM domains, considerable variation is seen in the overall length of members of the receptor superfamily. For example, the mature metabotropic glutamate receptor mGluR1a is composed of approximately 1180 amino acid residues[120] and receptors for the glycoprotein hormones follicle-stimulating hormone (FSH), luteinizing hormone/chorionic gonadotropin (LH/CG) and thyroid-stimulating hormone (TSH) range from approximately 670 to 740 residues in the mature receptor.[121] Adrenergic, dopamine, serotonin (5-hydroxytryptamine; 5-HT) and muscarinic acetylcholine receptors are typically on the order of 350 to 600 amino acids in length, as are many of the receptors which bind peptide ligands or lipid mediators.[122] Gustatory and odorant receptors, the visual color pigments and mammalian rhodopsins are generally smaller, ranging from 300 to 370 amino acid residues,[122,123] while the smallest G protein-coupled receptor described to date, the receptor for adrenocorticotropic hormone (ACTH), comprises only 297 residues.[124]

Variability in the overall length of receptors occurs primarily in the N-terminal extracellular domain. This is composed of 592 amino acid residues in the mature bovine parathyroid $Ca^{2+}$-sensing receptor (BoPCaR1),[125] between 550 and 575 residues in mature metabotropic glutamate receptors mGluR1-7[126-130] and 340 to 400 residues in the mature glycoprotein hormone receptors,[121] as compared to less than 10 residues in the adenosine $A_{2A}$ and $A_{2B}$ receptors.[131,132] Size variability is also seen in the third intracellular loop and the C-terminal intracellular domain (see Fig. 1.1f). The third intracellular loop is generally shorter in receptors which bind protein or peptide ligands or lipid mediators, relative to receptors for the bioactive amines. This segment ranges in size from 239 amino acids in the human m3 muscarinic acetylcholine receptor[133] to approximately 15 amino acids in the human complement C5a and N-formyl peptide receptors.[134-136] However, it is also very short, comprising only 13 amino acids, in the $Ca^{2+}$-sensing receptor BoPCaR1[125] and the metabotropic glutamate receptors mGluR1-7.[126-130] The C-terminal domain comprises 422 amino acids in the $Ca^{2+}$-sensing receptor BoPCaR1,[125] 359 amino acids in the metabotropic glutamate mGluR1a receptor,[120] 164 amino acids in the hamster $\alpha_{1B}$-adrenergic receptor,[137] 14 amino acids in the human $D_{2A}$ (long) and $D_{2B}$ (short) dopamine receptors[138] and is essentially absent, comprising only 2 amino acids, in the gonadotropin-releasing hormone (GnRH) receptor.[139]

The N-terminal extracellular domain, the third intracellular loop and the C-terminal cytoplasmic segment of the receptors also exhibit the greatest diversity in amino acid composition. There is no clear correlation between the size of these hydrophilic domains and the functional properties of ligand binding and G protein activation for any of the receptors, except for the glycoprotein hormone receptors, where the N-terminal extracellular domain has been demonstrated to comprise the high affinity ligand binding domain.[121,140,141] In these receptors, the large N-terminal domain is rich in Cys residues that may form disulfide bridges essential for maintenance of the three-dimensional structure of this extracellular region.[142]

In most receptors, the N-terminal domain contains from one to nine potential sites for Asn-linked (N-linked) glycosylation, represented by the sequence (Asn-X-Ser/Thr), where X is any amino acid except Pro or Asp.[143,144] Exceptions to this include the $A_1$, $A_{2A}$ and $A_{2B}$ adenosine receptors,[131,132,145] human and murine $\alpha_{2B}$ adrenergic receptors,[146,147] the rat serotonin 5-HT$_{2B}$ receptor,[148] and the human leukocyte platelet-activating factor (PAF) receptor.[149] However, the $A_1$, $A_{2A}$ and $A_{2B}$ adenosine and rat serotonin 5-HT$_{2B}$ receptors do exhibit potential sites for N-linked glycosylation in the second extracellular loop, as do a number of other receptors, including the dopamine $D_1$ and $D_5$, histamine $H_2$, IL-8 A and B, bradykinin $B_2$ and neuropeptide Y (NPY)/peptide YY (PYY) $Y_1$ receptor subtypes.[122] Potential N-linked glycosylation sites also occur in the first extracellular loop of some receptors, such as the human, ovine and rodent GnRH receptors,[139] the rat gastric inhibitory polypeptide (GIP) receptor,[150] the human melanocyte-stimulating hormone (MSH) MC-4[151] and avian purinergic $P_{2U}$ receptors,[152] and one of the known catfish odorant receptors.[153]

The glycoprotein nature of a number of G protein-coupled receptors, including adenosine, adrenergic, muscarinic acetylcholine, angiotensin II, somatostatin, LH/CG, TSH and $Ca^{2+}$-sensing receptors, as well as rhodopsin, has been demonstrated.[121,125,154-159] While it has not been established that all potential N-linked glycosylation sites occurring on G protein-coupled receptors are functional, the significance of N-linked glycosylation for receptor expression and ligand binding activity has been addressed for a number of receptors (see chapter 3).

Cell surface receptors usually contain an N-terminal hydrophobic leader sequence or signal peptide, which directs intracellular trafficking of newly synthesized receptors from the endoplasmic reticulum to the plasma membrane, and is cleaved from the receptor to generate the mature receptor protein.[160] Among G protein-coupled receptors, the existence of a signal peptide has been clearly demonstrated for the bovine endothelin ET$_B$ receptor, by comparison of the protein sequence encoded by the complementary deoxyribonucleic acid (cDNA) with the N-terminal sequence of the purified receptor.[161] Other receptors exhibit sequence features consistent with the presence of a leader sequence. Predictions of a signal peptide of 17 to 32 amino acids have been made for endothelin ET$_A$ and ET$_B$ receptor subtypes isolated from a variety of species,[162,163] as well as for the thrombin receptor,[164] the metabotropic glutamate receptors mGluR1-7,[126-130] the receptors for the glycoprotein hormones FSH, LH/CG and TSH,[121] the secretin receptor[165] and other receptors which are closely related to the secretin receptor at the level of primary amino acid sequence. These include the receptors for calcitonin,[166] corticotropin-releasing factor (CRF),[167-169] GIP,[150] glucagon and glucagon-like peptide 1 (GLP-1),[170,171] growth hormone-releasing hormone (GHRH),[172-174] parathyroid hormone/parathyroid hormone-related peptide (PTH/PTHrP),[175] pituitary adenylyl cyclase-activating peptide (PACAP)[176] and vasoactive intestinal peptide (VIP).[177]

On the other hand, the absence of a signal peptide has been demonstrated unambiguously for the murine bombesin/gastrin-releasing peptide (GRP) BB$_2$ receptor, through correlation of the protein sequence encoded by the cDNA with the N-terminal sequence of the purified receptor.[178] This receptor is representative of the majority of G protein-coupled receptors described to date, which do not display the stretch of hydrophobic residues characteristic of a signal peptide in their N-terminal domain. The mechanism whereby these receptors are localized to the cell surface remains unknown.

General structural and functional features of G protein-coupled receptors are summarized in Figure 1.4. For adrenergic and muscarinic acetylcholine receptors, the binding site for ligands has been localized to a pocket buried within the lipid bilayer and formed by contributions from amino acid side chains in several of the hydrophobic helices of the receptor. This is analogous to the binding of 11-*cis*-retinal to rhodopsin, which involves the formation of a Schiff base with the amine side chain of a Lys residue in TM7.[179] Similar principles probably apply to the binding of the biogenic amines dopamine, histamine and serotonin, and of other small ligands such as adenosine, to their respective receptors.[180] An exception is seen in the binding of the amino acid glutamate to metabotropic glutamate receptors. In these receptors, the large N-terminal extracellular domain exhibits ligand recognition properties and constitutes the determinant for receptor subtype selectivity.[181] The receptors for the multimeric glycoprotein hormones FSH, LH/CG and TSH also exhibit a large N-terminal extracellular domain, which has been identified as the high affinity binding site.[182] Smaller peptide ligands, on the other hand, may interact with multiple receptor domains. For example, in receptors for the tackykinin peptides substance P, neurokinin A (NKA; substance K) and neurokinin B (NKB; neuromedin K), several extracellular and TM segments contribute to the ligand binding site and in provision of determinants for receptor subtype selectivity.[183] Structural determinants for ligand binding and recognition which have been identified for a number of G protein-coupled receptors are discussed in detail in chapter 3.

Mutagenesis studies on adrenergic, muscarinic acetylcholine and glycoprotein hormone receptors have established that all of the intracellular domains are involved in efficient functional coupling of receptors to G proteins. Receptor-G protein interaction relies critically on the third cytoplasmic loop and of particular importance in this regard are the segments of this loop immediately adjacent to TM5 and TM6. There is little primary sequence homology within these domains and it is their secondary structure that is thought to be important in mediating interactions with G proteins. Additional sites of importance in G protein interaction are the first and second cytoplasmic loops and the N-terminal portion of the C-terminal tail of the receptors (see chapter 3).[1,157,182,184-190]

The C-terminal cytoplasmic domain is usually rich in Ser and Thr residues that are potential sites for phosphorylation by kinases such as the β-adrenergic receptor kinase (βARK).[187] Both the third intracellular loop and the C-terminal domain usually bear potential sites for phosphorylation by cAMP-dependent kinase (protein kinase A; PKA) or protein kinase C (PKC). The functional significance of these phosphorylation sites in receptor desensitization has been clearly demonstrated in the β-adrenergic receptors (see chapter 3).[188,190,191]

Almost all G protein-coupled receptors contain Cys residues in the first and second extracellular loops. Mutation of these Cys residues results in altered function of rhodopsin, β$_2$-adrenergic and muscarinic

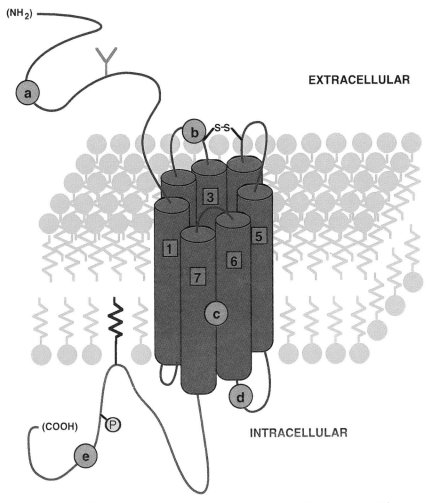

Fig. 1.4. Structural and functional domains of G protein-coupled receptors. Schematic representation of a G protein-coupled receptor with putative transmembrane domains depicted as cylinders. Letters refer to structural domains to which some general, but not necessarily universal, functional features have been assigned. (a) The N-terminal extracellular domain contains potential sites for N-linked glycosylation (**Y**) in most receptors. It constitutes the ligand binding domain for glycoprotein hormone and metabotropic glutamate receptors and contributes ligand binding determinants in a number of receptors that bind peptide ligands. (b) The first and second extracellular loops contain Cys residues that are involved in disulfide bond formation and are of importance in maintenance of the structural integrity of adrenergic and muscarinic acetylcholine receptors and of rhodopsin. (c) The transmembrane domains contain residues critical for ligand binding in adrenergic and muscarinic acetylcholine receptors, and for binding of 11-cis-retinal to rhodopsin. (d) The intracellular loops and the N-terminal part of the C-terminal cytoplasmic domain are involved in coupling to G proteins, with particular importance ascribed to the segments of the third intracellular loop immediately adjacent to transmembrane domains 5 and 6. (e) The C-terminal cytoplasmic domain contains a Cys residue that represents the site for palmitoylation of receptors and attachment to the lipid bilayer. Both the C-terminal cytoplasmic domain and the third intracellular loop contain Ser and Thr residues that are phosphorylated during receptor desensitization.

acetylcholine receptors.[192-196] These data are consistent with the existence of intramolecular disulfide bond(s) between the adjacent first and second extracellular loops in native receptors which are critical for maintaining the tertiary structure of the receptor proteins.

Most receptors also exhibit conservation of a Cys residue in the N-terminal segment of the cytoplasmic C-terminal domain. This is the site of palmitoylation in rhodopsin and the $\beta_2$-adrenergic receptor.[197,198] This post-translational modification would result in the anchoring of receptors to the membrane, thereby introducing an additional intracellular loop into the molecular architecture of the receptor (see Fig. 1.4).

The greatest degree of amino acid similarity amongst G protein-coupled receptors occurs in the TM segments. The sequence similarity varies from 20-90%, with greatest conservation of sequence observed between subtypes of a specific receptor.[122,199] There are a number of sequence features which are generally conserved amongst most members of the G protein-coupled receptor superfamily. These are represented schematically in Figure 1.5, within the framework of the human $\beta_2$-adrenergic receptor.[200,201]

Particularly well conserved among most members of the receptor superfamily are several Pro residues, occurring in TM4, 5, 6 and 7. These residues are thought to introduce kinks in the α-helices and may be important in formation of the ligand binding pocket for receptors such as the muscarinic acetylcholine receptors, which bind ligand in the hydrophobic core of the protein (see chapter 3). In addition, the Pro residue in TM7 of the hamster $\beta_2$-adrenergic receptor[202] and Pro residues in TM5, TM6 and TM7 of the rat m3 muscarinic acetylcholine receptor[203] have been shown to be important for expression of receptors at the cell surface. The structural role of Pro residues in maintenance of correct protein folding may be of significance in efficient intracellular trafficking of nascent receptors or their stable integration into the plasma membrane. Other residues well conserved in

most members of the receptor superfamily are (Gly, Asn and Val) residues in TM1; (Leu, Ala, Ala and Asp) residues in TM2; (Trp) in TM4; (Phe and Trp) in TM6; and (Asn and Tyr) in TM7.

Certain conserved residues are replaced in particular subfamilies. For example, an Asp residue which occurs in TM3, and corresponds to Asp113 in the $\beta_2$-adrenergic receptor (see Fig. 1.5), is conserved in adrenergic, muscarinic acetylcholine, dopamine, histamine and serotonin receptors. It plays a role in ligand binding and receptor activation in these receptors (see chapter 3). However, it is not found in adenosine receptors or in many receptors which bind peptide ligands. For example, it is replaced by Phe in the glycoprotein hormone receptors and by Pro in the tachykinin peptide receptors. Similarly, the conserved Trp in TM6 is replaced by Met in the glycoprotein hormone receptors, and Asp replaces Asn in TM7 of receptors for the prostanoid prostaglandin E (PGE$_2$).[122,204] The sequence (Asp-Arg-Tyr), which occurs on the intracellular face of TM3, is very highly conserved, with the Arg being invariant and the Asp and Tyr conservatively replaced in several receptors. A role for the conserved

*Fig. 1.5. (opposite) Model of the human $\beta_2$adrenergic receptor. Putative transmembrane domains are defined and numbered consecutively. Amino acid residues shown in blue are conserved in the majority of G protein-coupled receptors. Disulfide bond formation (−S-S−) between Cys106/Cys191 and Cys184/Cys190, and palmitoylation involving Cys341, are shown. Potential sites for N-linked glycosylation are identified (Y). Potential sites for phosphorylation (P) comprise sites for phosphorylation by protein kinase A (Ser262 within the third cytoplasmic loop and Ser346 within the N-terminal segment of the C-terminal cytoplasmic domain), and potential sites for phosphorylation by β-adrenergic receptor kinase, which correspond to Ser and Thr residues extending from Ser355 within the C-terminal cytoplasmic domain. [Reprinted with permission from DNA and Cell Biology Volume 11, Probst WC, Snyder LA, Schuster DI et al, Sequence alignment of the G protein-coupled receptor superfamily, Pages 1-20, Copyright (1992); and from the Annual Review of Neuroscience Volume 15, © 1992, by Annual Reviews Inc.[188]].*

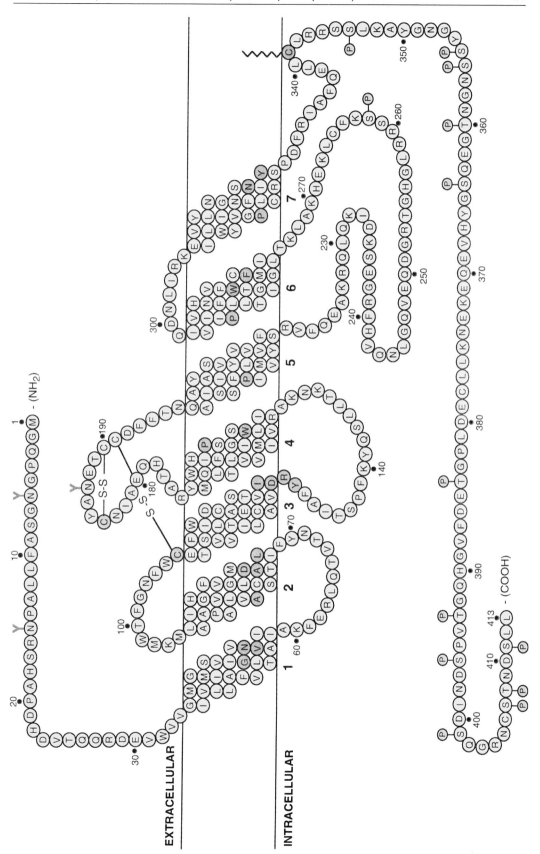

Asp/Glu residue in G protein coupling has been described for rhodopsin, adrenergic, muscarinic acetylcholine and angiotensin II receptors (see chapters 3 and 4), while it may be of significance in correct intracellular processing and expression of LH/CG receptors.[205]

## MOLECULAR CLONING OF G PROTEIN-COUPLED RECEPTORS

Two major strategies have been used for the cloning of G protein-coupled receptors.

The first strategy involves the use of ligand binding specificity in various applications, including classical approaches which involve receptor purification and sequencing, as well as a variety of expression cloning approaches which rely upon detection of either ligand binding or the functional activation of heterologously expressed receptors.

The second strategy for cloning of receptor sequences has incorporated a number of nucleotide homology screening approaches which were based upon observations and predictions of conservation of primary amino acid sequence. Such sequence conservation is seen in the broad context of sequence features which are generally characteristic of members of the receptor superfamily, as discussed in the preceding section (see Fig. 1.5), and is also evident at the level of receptors interacting with the same ligand, both within and between species. This second cloning strategy has resulted in the rapid isolation of a large number of receptors and receptor subtypes, as a consequence of the extensive sequence homology now known to exist between members of the superfamily. It has allowed the identification of molecular subtypes for particular ligands which were not delineated by previous pharmacological characterizations of receptor populations. It has been particularly fruitful in the isolation of novel receptor sequences and has allowed expansion of existing definitions of endogenous ligands, through the isolation of discrete receptors for which the bioactive molecule has been identified subsequently.

## 1. LIGAND BINDING SPECIFICITY AND EXPRESSION CLONING

Included in this category of approaches for receptor cloning is the classical methodology which involves receptor purification followed by screening of cDNA or genomic DNA libraries with oligodeoxynucleotide probes based on primary amino acid sequence. This approach was used for molecular cloning of the hamster $\beta_2$-adrenergic receptor[99] and has been applied successfully in the case of other receptors such as the human oxytocin receptor[206] and the murine bombesin/GRP BB$_2$ receptor.[178]

Expression cloning procedures involving the functional expression of cDNAs in heterologous cells or cell lines also rely upon specific ligand interaction with the receptor of interest. A variety of expression systems and detection procedures have been developed and implemented successfully. The use of *Xenopus* oocytes for functional expression of receptor cDNAs involves the injection of messenger ribonucleic acid (mRNA) or complementary RNA (cRNA) into oocytes. Activation of expressed receptors by binding of specific ligand triggers phosphatidylinositol (PI) hydrolysis, release of $Ca^{2+}$ from intracellular stores and opening of oocyte membrane $Cl^-$ channels, generating currents in voltage-clamped oocytes which can be recorded electrophysiologically. This approach was pioneered by Nakanishi and coworkers in cloning the bovine NK$_2$ (NKA; substance K) receptor[207] and has been used subsequently for cloning of receptors such as the bovine endothelin ET$_A$ receptor, the guinea pig PAF receptor,[208] the murine thyrotropin-releasing hormone (TRH) receptor and the rat bradykinin B$_2$ receptor.[204]

A large number of receptors have also been cloned through functional expression in mammalian cell lines. The most widely used mammalian expression cloning system incorporates COS-1 or COS-7 cells transiently transfected with plasmid DNA prepared from cDNA libraries constructed in mammalian expression vectors carrying the SV40 origin of replication.[209,210] The detection of heterologously expressed receptors relies upon ligand binding. A preliminary enrichment for the clones of interest may be achieved by an immunoselective procedure termed "ligand panning",[211] using monoclonal antibodies directed against the ligand[212] or against molecular "tags", such as biotin, incorporated into the ligand molecule.[213] Whether or not ligand panning is used, individual clones are isolated through successive rounds of transfection and screening. Generally, radioligands are used for screening and cell-associated radioactivity is discriminated by direct counting, image analysis or autoradiography. Examples of G protein-coupled receptors cloned using this approach and incorporating some of the possible variations of procedure described above include the murine δ-opioid receptor, rat angiotensin II $AT_1$ and secretin receptors, human IL-8 A receptor and the porcine calcitonin C1 receptor.[204,214-216]

Expression cloning and screening procedures incorporating bioassays for activation of intracellular pathways in response to ligand-mediated activation of heterologously expressed receptors have also been described. For example, the human vasopressin $V_2$ receptor cDNA was cloned using a procedure which relied upon assaying for acquisition of vasopressin responsiveness in the stimulation of adenylyl cyclase in mouse fibroblast Ltk⁻cells transfected with the selectable marker thymidine kinase and co-transfected with human genomic DNA. Transformants were generated which contained successively less human genomic DNA but retained responsiveness to vasopressin. Construction of a genomic DNA library from one such transformant afforded isolation of the human vasopressin $V_2$ receptor gene and its functional expression in stably transfected mouse cells provided the means for isolation of the cDNA encoding this human receptor.[217] Another method which takes advantage of elevation of intracellular cAMP levels by stimulation of adenylyl cyclase through activated receptors is based on transcriptional induction of a cAMP-responsive luciferase reporter gene. The cDNA encoding the rat PACAP type-I (PACAPR-2) receptor was cloned

using this procedure by expression of the appropriate reporter gene construct in porcine renal epithelial LLC PK1 cells.[176] This methodology could be used to screen for a variety of receptors which couple to stimulation of adenylyl cyclase activity.

*Xenopus laevis* melanophores are capable of functional expression of recombinant receptors which couple to either stimulation or inhibition of adenylyl cyclase or the stimulation of phospholipase C. Receptor-mediated stimulation of either of these enzymes results in darkening of melanophores, due to cytoplasmic dispersion of melanosomes which contain the dark pigment melanin. Inhibition of cAMP accumulation by receptor-mediated inhibition of adenylyl cyclase induces aggregation of melanosomes towards the cell center, with consequent lightening of the cell. Monitoring of pigment translocation by video imaging provides a means for visualization of functional receptor activation at single cell resolution. The functional activation of recombinant murine bombesin/GRP $BB_2$ and rat substance P ($NK_1$) receptors, which couple to stimulation of phospholipase C, and of human $β_2$-adrenergic and dopamine $D_2$ receptors, which couple to the stimulation and inhibition of adenylyl cyclase, respectively, has been demonstrated in transfectants of a clonal cell line isolated from primary cultures of melanophores.[218] This is a novel and versatile bioassay which not only allows functional activation of cloned receptors to be analyzed, but also represents a potential means for screening cDNA libraries to allow clones encoding new receptors to be isolated.

## 2. HOMOLOGY CLONING

The conservation of amino acid sequence features seen among members of the G protein-coupled receptor superfamily is reflected at the DNA level and has been used for the isolation of a large number of additional members of the G protein-coupled receptor superfamily. This approach involves reduced stringency hybridization screening of cDNA and genomic DNA libraries using cDNA or oligodeoxynucleotide probes, or by the use of polymerase chain reaction (PCR)-based cloning approaches.

Reduced stringency hybridization screening has allowed isolation of homologs of particular receptor subtypes across species. For example, the human bombesin/GRP $BB_2$ and bombesin/neuromedin B $BB_1$ receptors were isolated using murine bombesin/GRP $BB_2$ and rat bombesin/neuromedin B $BB_1$ receptors, respectively, as hybridization probes, and the rat NPY/PYY $Y_1$ receptor was used to isolate the cDNA encoding the human NPY/PYY $Y_1$ receptor. Similarly, hybridization screening with a segment of the porcine m2 muscarinic acetylcholine receptor allowed isolation of the human m2 muscarinic acetylcholine receptor and the human endothelin $ET_A$ receptor was isolated using the bovine receptor cDNA sequence.[133,204,219] Oligodeoxynucleotide probes have also been used as, for example, in the isolation of the human $D_{2A}$ (long) dopamine receptor using a mixed probe of 45 nucleotides based on a segment of the sequence of TM2 of the rat $D_{2B}$ (short) dopamine receptor,[138] and isolation of the human adenosine $A_1$ receptor using oligodeoxynucleotide probes corresponding to segments of the second extracellular and third intracellular domains of the canine adenosine $A_1$ receptor.[145] The success of this approach reflects low interspecies variability among members of the receptor superfamily. While the complement C5a receptor shows only 65-75% identity at the amino acid level, for the majority of receptors, between 85-99% identity is generally observed for a particular receptor subtype across mammalian species.[220]

Additional molecular subtypes of receptor for a particular ligand within and between species have also been isolated using low stringency hybridization screening. Thus, the human dopamine $D_5$ receptor subtype was cloned using a segment of the human dopamine $D_1$ receptor cDNA,[221] and human m1, m3 and m4 muscarinic acetylcholine receptor subtypes were isolated using a segment of the porcine m2 muscarinic acetylcholine receptor[133] Amino

acid identity between molecular subtypes within a given species is generally on the order of 40-60%.[133]

Low stringency hybridization screening has also identified receptors which are fairly closely related at the amino acid sequence level but which bind a ligand different from that which is specific for the cloned receptor sequence used for screening. For example, the rat receptor for VIP was isolated using rat secretin receptor cDNA as a hybridization probe. These two receptors exhibit 48% identity at the amino acid level.[222] The isolation of the rat $D_{2B}$ (short) dopamine receptor using a segment of the hamster $\beta_2$-adrenergic receptor reflects 39% identity at the amino acid level between these two discrete receptors and represents an example of the utility of this technique for the isolation of novel receptors across species.[223] Oligodeoxynucleotide probes have also been employed, such as in the use of an oligodeoxynucleotide corresponding to 56 bases spanning TM2 of the rat $NK_2$ (NKA; substance K) receptor sequence to isolate a rat receptor which shows functional activation in response to cannabinoids.[224] The ensuing search for an endogenous ligand for this receptor resulted in the identification of arachidonylethanolamide, or anandamide, a derivative of arachidonic acid, as an endogenous cannabimimetic.[225,226]

Amino acid sequence features generally conserved throughout the G protein-coupled receptor superfamily, such as those highlighted in Figure 1.5, have been used by many investigators to design degenerate oligodeoxynucleotide primers for PCR amplification of receptor sequences from both cDNA and genomic DNA templates. This approach has led to the isolation of a large number of receptors, including canine and human adenosine $A_1$ and $A_{2A}$ receptor subtypes,[131,227-229] somatostatin $SRIF_{1A}$ and $SRIF_{2A}$ receptor subtypes,[204,230] and rat and human receptors for odorant molecules.[231,232] However, some of the novel receptor sequences isolated in this manner encode receptors for which the ligand remains undefined. These so-called "orphan" receptors include two human sequences which are closely related to the human $N$-formyl peptide receptor,[233,234] human, murine and rat sequences which bear substantial homology with human IL-8 A and B receptors,[235-237] and a rat sequence which is most closely related to the canine orphan receptor RDC1, that was itself originally also isolated using this methodology.[227,237,238]

PCR amplification procedures have also been used, in a manner analogous to low stringency hybridization screening of libraries, for the isolation of receptor sequences across species and of additional molecular subtypes of cloned receptors. This more directed approach allowed, for example, the isolation of the MC-4 subtype of the human MSH receptor. PCR amplification primers were based on the sequence of MC-1 and MC-3 MSH receptor subtypes, as well as the closely related ACTH receptor.[151] Similarly, the rat $\alpha_{1D}$-adrenergic receptor subtype was isolated using PCR amplification primers based on the hamster $\alpha_{1B}$ or bovine $\alpha_{1C}$-adrenergic receptor subtypes, and incorporation of solution phase hybridization screening of a rat hippocampal cDNA library for isolation of a full-length receptor sequence.[239] Novel receptors have also been isolated in the search for additional molecular subtypes of existing receptors. For example, degenerate oligodeoxynucleotide primers based on the sequence of somatostatin $SRIF_{1A}$ and $SRIF_{2A}$ receptor subtypes afforded isolation of the murine $\delta$ and $\kappa$ opioid receptors.[240]

# MEMBERS OF THE G PROTEIN-COUPLED RECEPTOR SUPERFAMILY

## 1. CLONED RECEPTORS AND RECEPTOR SUBTYPES

Molecular cloning approaches have resulted in the isolation of structurally and functionally distinct subtypes of receptor for a large number of endogenous ligands, allowing for considerable extension of existing pharmacological definitions of receptor subtypes for known bioactive molecules and contributing greatly to our understanding of receptor diversity. For example, prior to the molecular cloning and characterization of dopamine receptors, two pharmacologically discrete subtypes of dopamine receptor, $D_1$ and $D_2$, had been identified in mammalian tissues. Molecular characterization has now clearly identified seven molecular subtypes of receptor which interact with dopamine, designated $D_1$, $D_{2A}$ (long), $D_{2B}$ (short), $D_3$ (long), $D_3$ (short), $D_4$ and $D_5$. In addition, polymorphic variants have been described for the $D_2$ and $D_4$ dopamine receptor subtypes. The $D_{2A}$ (long) and $D_{2B}$ (short), $D_3$ and $D_4$ dopamine receptor subtypes have a $D_2$-like pharmacological profile, whereas the $D_1$ and $D_5$ dopamine receptor subtypes exhibit $D_1$-like pharmacology.[241-244]

The cloning and expression of receptors in heterologous systems has also allowed the functional capabilities of receptor subtypes to be examined. Such analyses have demonstrated that the existence of distinct molecular subtypes of receptor is responsible, at least in part, for the diversity of function elicited by individual ligands. Thus, the $D_1$ and $D_5$ dopamine receptor subtypes have been shown to couple to stimulation of adenylyl cyclase activity, while $D_2$ and $D_4$ dopamine receptor subtypes couple to inhibition of adenylyl cyclase activity.[230] However, the two molecular variants of the $D_2$ dopamine receptor, $D_{2A}$ (long) and $D_{2B}$ (short), which are indistinguishable from one another according to pharmacological criteria, have been shown to be functionally distinct in respect of the maximal inhibition of adenylyl cyclase activity which they elicit. These molecular variants of the $D_2$ dopamine receptor are generated by alternative splicing of the primary mRNA transcript encoding them and differ from one another by virtue of a 29 amino acid insertion within the third intracellular loop of the receptor.[138,245-249] Differential signal transduction properties have also been attributed to splice variants of the rat PACAP type-I (PACAPR-2) receptor which differ in the third intracellular loop,[176] while certain splice variants of the murine $PGE_2$ $EP_3$

receptor, which differ in the C-terminal tail of the receptor, exhibit differential G protein specificity.[250,251] These findings highlight the importance of the third extracellular loop and C-terminal domain of G protein-coupled receptors in G protein interactions (see chapter 3).

The range of receptors for which splice variants have now been identified will be described in Section 3 (ii) (below) and the physiological significance of receptor subtype diversity will be discussed in Section 4. The relevance to health and disease, and therapeutic potential associated with description of molecular subtypes of specific receptors, will be addressed in chapter 4.

Cloned receptor sequences described to date are listed in Table 1.1. This listing highlights the broad range of bioactive molecules which interact with this class of receptor, ranging from photons of light which activate the visual color pigments and rhodopsin, inorganic ions such as $Ca^{2+}$, amino acids such as glutamate, the catecholamine neurotransmitters including noradrenaline and dopamine, the purine nucleoside adenosine and purine nucleotides cAMP and ATP, lipid mediators such as PAF and the prostanoids $PGE_2$ and thromboxane $A_2/PGH_2$, peptides including the tripeptide TRH and the 36 amino acid peptide NPY, and multi-subunit proteins exemplified by the glycoprotein hormones FSH, LH/CG and TSH.

Receptors and pharmacologically defined receptor subtypes known to mediate cellular responses by interaction with G proteins which have not yet been cloned include receptors for bradykinin ($B_1$), calcitonin gene-related peptide, galanin, GABA ($GABA_B$), histamine ($H_3$), serotonin ($5-HT_4$), leukotrienes ($LTB_4$, $LTC_4$ and $LTD_4$), melatonin, prostaglandin $PGD_2$ (DP) and vasopressin ($V_{1B}$).

## 2. RECEPTOR SUBFAMILIES

Members of the G protein-coupled receptor superfamily may now be categorized into discrete subfamilies according to conservation of distinguishing amino acid sequence motifs and overall structural organization.

The majority of cloned receptors exhibit most of the amino acid sequence features highlighted in Figure 1.5.[122] However, within this broad grouping, the gustatory and odorant receptors constitute two large families, whose members are more closely related to one another than they are to other G protein-coupled receptors and within which receptors may be grouped according to sequence similarity.[123,153,231,232,252-254] Similarly, the glycoprotein hormone receptors represent a discrete subfamily with distinguishing sequence features. Their characteristically large N-terminal domain contains between 9 and 14 copies of an imperfectly repeated sequence of approximately 25 amino acids similar to a repeated motif called a "leucine-rich repeat". The leucine-rich repeat is characteristic of a diverse group of serum and membrane glycoproteins known collectively as the leucine-rich glycoproteins (LRG), which includes the human serum protein $\alpha_2$-glycoprotein, the human platelet membrane glycoprotein 1b, yeast adenylyl cyclase and a *Drosophila* membrane glycoprotein involved in photoreceptor cell morphogenesis. The functional significance of the leucine-rich repeat structure is not clear. However, these repeats are likely to form amphipathic helices or β-sheets which could interact with both hydrophilic and hydrophobic surfaces.[122,140,141,255]

The metabotropic glutamate receptors mGluR1-7 show a common structural architecture with other members of the G protein-coupled receptor superfamily, comprising a large extracellular N-terminal domain that precedes 7 TM segments and an intracellular C-terminal domain.[126-130] They are closely related to one another, with greatest similarity observed between mGluR1 and 5, mGluR2 and 3, and mGluR4, 6 and 7. However, sequence homology is observed with only one other member of the G protein-coupled receptor superfamily described to date, the bovine $Ca^{2+}$-sensing receptor BoPCaR1, which is most closely related to mGluR1 and 5.[125] Thus, the metabotropic glutamate receptors and the $Ca^{2+}$-sensing receptor define a novel subfamily of the G protein-coupled receptor superfamily. Sequence features conserved among members of this receptor subfamily are shown in Figure 1.6, within the framework of the $Ca^{2+}$-sensing receptor BoPCaR1.

Another discrete receptor subfamily has been defined with the isolation of receptors for secretin, calcitonin, CRF, GIP, glucagon, GLP-1, GHRH, PTH/PTHrP, PACAP and VIP, as well as for the diuretic hormone of the tobacco hornworm *Manduca sexta*. Distinguishing sequence features of this receptor subfamily are shown in Figure 1.7, within the framework of the rat secretin receptor. All known members of this receptor subfamily couple to the stimulation of adenylyl cyclase activity, while calcitonin, glucagon, GLP-1, PACAP and PTH/PTHrP receptors couple also to PI metabolism and the mobilization of intracellular $Ca^{2+}$.[167,171,176,230,256-260]

## 3. EVOLUTIONARY CONSIDERATIONS

The molecular cloning of a large number of G protein-coupled receptors, with the identification of discrete subfamilies and elucidation of gene structure and chromosomal localization in many instances, have greatly increased our understanding of evolutionary mechanisms responsible for generating the complex multiplicity of members of this receptor superfamily.

The existence of a common ancestral receptor molecule, with divergence attributable to successive gene duplications, is suggested by the striking conservation of sequence features by the majority of G protein-coupled receptors, particularly in TM regions, where tertiary structure is crucial for the generation of functional receptors, and in the spacing of key functional amino acid residues (see Figures 1.4 and 1.5). Not surprisingly, the greatest sequence homology and structural conservation occurs among receptors which interact with identical or closely related ligands.[122] Members of the metabotropic glutamate and secretin receptor subfamilies have a similar overall structure that incorporates 7 TM domains with extracellular N-terminal and cytoplasmic C-terminal segments, and functional activation relying on interaction with intracellular G proteins. However, members of these subfamilies exhibit no sequence similarity with the majority of G protein-coupled receptors. The evolutionary relationship of these receptor subfamilies to other members of the superfamily is unclear and raises the possibility of convergent evolution. Nevertheless, the identification of distinguishing sequence features for the metabotropic glutamate and secretin receptor subfamilies (see Figures 1.6 and 1.7) is consistent with gene duplication events contributing to the generation of structural diversity in these subfamilies also.

Nucleotide and protein sequence analysis, as well as analysis of gene structure, have been used in efforts to elucidate the time frame and mechanisms of evolution of G protein-coupled receptors and receptor subtypes. The direct alignment of the sequence of the prokaryotic membrane protein bacteriorhodopsin with G protein-coupled receptor sequences gives no indication of significant primary sequence homology. However, if the sequential order of TM helices in bacteriorhodopsin is ignored, considerable homology is observed between TM helices III, VII and I of bacteriorhodopsin with TM helices 5, 3 and 7, respectively, of G protein-coupled receptors. This has led to the suggestion that exon shuffling may have occurred in evolution of the ancestral G protein-coupled receptor gene from the gene encoding bacteriorhodopsin.[110] An alternative hypothesis is based upon the observation of intragenic as well as intergenic similarities between helices 1-3 and 5-7 of bacteriorhodopsin and a number of adrenergic and muscarinic acetylcholine G protein-coupled receptors, leading to the suggestion that an ancestral gene may have evolved through a duplication event, with TM helices 5-7 originating as duplicates of helices 1-3.[261] This hypothesis is supported by conservation of the 11-*cis*-retinal-binding Lys residue in TM7 of both bacteriorhodopsin and the mammalian visual color pigments.[122,179,262]

## Table 1.1. Cloned G protein-coupled receptors and receptor subtypes

| | | Reference* |
|---|---|---|
| **PEPTIDES AND PEPTIDE HORMONES** | | |
| Adrenocorticotropic hormone | | 347 |
| Angiotensin II | $AT_{1A}$, $AT_{1B}$, $AT_{1C}$, $AT_2$, $AT_3$; mas | 291,340,348-354 |
| Bombesin/gastrin-releasing peptide | $BB_2$, $BB_3$ | |
| Bombesin/neuromedin B | $BB_1$ | |
| Bradykinin | $B_2$ | |
| • • Calcitonin | C1a[a], C1b[a] | 166,256,355 |
| Cholecystokinin/gastrin | $CCK_A$, $CCK_{B1}$[a], $CCK_{B2}$[a] | 274,356 |
| Complement C5a | | 135,136,220 |
| • • Corticotropin-releasing factor | (2)[a] | 167-169 |
| • • Diuretic hormone | | |
|     (Tobacco hornworm *Manduca sexta*) | | 260 |
| Endothelin | $ET_A$, $ET_B$, $ET_C$ | 163,357 |
| Follicle-stimulating hormone | | 121,358,359 |
| • • Gastric inhibitory polypeptide | | 150 |
| • • Glucagon | | 170,257,360 |
| • • Glucagon-like peptide 1 | | 171,361,362 |
| Gonadotropin-releasing hormone | | 139,363-367 |
| • • Growth hormone releasing hormone | (3)[a] | 174 |
| Interleukin-8 | A, B | |
| Luteinizing hormone/chorionic | | |
|     gonadotropin | | 121,141 |
| Macrophage inflammatory protein | | |
|     1α/RANTES | | |
| Monocyte chemoattractant protein 1 | A[a], B[a] | 304 |
| Melanocyte-stimulating hormone | MC-1, MC-2, MC-3, MC-4 | 151,347,368-370 |
| N-formyl peptide | (2) | 134 |
| Neurokinin A (substance K) | $NK_2$ | 371 |
| Neurokinin B (neuromedin K) | $NK_3$ | |
| Neuropeptide Y/peptide YY | $Y_1$, $Y_2$ | 372 |
| Neurotensin | | |
| Opioids | δ, κ, μ | 212,373 |
| • • Parathyroid hormone/parathyroid | | |
|     hormone-related peptide | | 175,258,374-376 |
| • • Pituitary adenylyl cyclase activating | | |
|     peptide | 1, 2 (I[a], I-hip[a], I-hop1[a], I-hop2[a], I-hip-hop1[a]), 3 | |
| • • Secretin | | |
| Somatostatin | $SRIF_{1A}$ (2)[a], $SRIF_{1B}$, $SRIF_{1C}$, $SRIF_{2A}$, $SRIF_{2B}$ | 377 |
| Substance P | $NK_{1A}$[a], $NK_{1B}$[a] | |
| Thrombin | | 164,378,379 |
| Thyroid-stimulating hormone | | 121 |
| Thyrotropin-releasing hormone | (2)[a] | 303,380-387 |
| • • Vasoactive intestinal peptide | $VIP_1$, $VIP_2$ | |
| Vasopressin/oxytocin | $V_{1A}$, $V_2$, OT | 388,389 |
| **NEUROTRANSMITTERS** | | |
| Adenosine | $A_1$, $A_{2A}$, $A_{2B}$, $A_3$ | |
| Adrenergic | $α_{1B}$, $α_{1C}$, $α_{1D}$, $α_{2A}$, $α_{2B}$, $α_{2C}$, $β_1$, $β_2$, $β_3$ | 227,281,390-393 |
| Dopamine | $D_1$, $D_{2A}$[a], $D_{2B}$[a], $D_3$ (2), $D_4$[b], $D_5$ | 394,395 |
| • Glutamate | mGluR1a,b,c[a], mGluR2, mGluR3, | |
| | mGluR4, mGluR5, mGluR6, mGluR7 | |
| Histamine | H1, H2 | 396 |
| Muscarinic acetylcholine | m1, m2, m3, m4, m5 | 397-400 |
| Octopamine/tyramine | | 401,402 |
| Serotonin (5-hydroxytryptamine) | $5\text{-}HT_{1A}$, $5\text{-}HT_{1B}$, $5\text{-}HT_{1Dα}$, $5\text{-}HT_{1Dβ}$, $5\text{-}HT_{1E}$, $5\text{-}HT_{1F}$, | |
| | $5\text{-}HT_{2A}$, $5\text{-}HT_{2B}$, $5\text{-}HT_{2C}$, | |
| | $5\text{-}HT_{5A}$, $5\text{-}HT_{5B}$, $5\text{-}HT_6$, $5\text{-}HT_{7A}$, $5\text{-}HT_{7B}$ | 227,284,403-409 |
| **INORGANIC IONS** | | |
| • $Ca^{2+}$ | | 125,410 |

## Table 1.1. continued

| | | Reference* |
|---|---|---|
| **LIPID MEDIATORS** | | |
| Anandamide | (2) | |
| Platelet-activating factor | | 149,208,411,412 |
| Prostacyclin (PGI$_2$) | IP | 413 |
| Prostaglandin E (PGE$_2$) | EP$_1$, EP$_2$, EP$_3$ (4)[a] | 251,301,305,414-418 |
| Prostaglandin F (PGF$_2\alpha$) | FP | 419-421 |
| Thromboxane A$_2$/prostaglandin H$_2$ | TP | 422 |
| **SENSORY STIMULI** | | |
| Gustatory | (>60) | 123,254 |
| Light | blue, green and red visual color pigments, rhodopsin | 122 |
| Odorants | (>100) | 153,231,232, 252,253 |
| **OTHER REGULATORY FACTORS** | | |
| cAMP (*Dictyostelium discoideum*) | CAR1, CAR2, CAR3 | 264,423 |
| ATP | P$_{2Y1}$, P$_{2U}$ | 424 |
| Tachykinin-like peptides | DTKR, NKD | 425,426 |
| Yeast mating factors | STE-2, STE-3 | 265-267,427 |
| **ORPHAN RECEPTORS** | | |
| 4-24 (G10d) | | 237,238 |
| 6H1 | | 428 |
| • • Calcitonin-receptor-like | Group 1 | 429 |
| | Group 2 | 429 |
| AGR16 | | 430 |
| BLR1 | | 431 |
| EBI 1 | | 432 |
| EBI 2 | | 432 |
| edg-1 | | 433 |
| FPRH1 | | 233 |
| FPRH2 | | 233 |
| FPRL1 (RFP, FPR2) | | 234,434,435 |
| Ghra | | 436 |
| GIR | | 437 |
| GPCR01[c], R334[c] | | 438,439 |
| HNB7,[c]GPCR21,[c] R4[c] | | 437a,438 |
| LCR1[c], D2S201E (L5, hFB22, LESTR)[c] | | 236,333,334,337-339 |
| *mrg* | | 342 |
| NLR | | 440 |
| RDC1 | | 227,441,442 |
| ROR-C | | 443 |
| RTA | | 341 |
| Cytomegalovirus | US27, US28, US33 | 444,445 |
| Herpesvirus saimiri | ECRF3 | |
| Swinepox virus | K2R | 446 |

Receptors are categorized according to endogenous ligand. Receptor subtype designations are based primarily on the nomenclature tabulation of Watson and Girdlestone (1994).[230] For receptors whose nomenclature has not yet been finalized, the number of subtypes is given in parentheses.

\* Citations for cloned receptor sequences are available in PIR protein database entries as listed in Watson and Girdlestone (1994),[230] and only literature references in addition to those available from this source are listed in this Table.

[a] These receptor subtypes result from alternative splicing of a single mRNA transcript.

[b] Two distinct types of polymorphic variant of the dopamine D$_4$ receptor have been described. The first type carries an additional four amino acids in the N-terminal extracellular domain, while the second type comprises variants containing two, three, four, five and seven-fold repeats of a 16 amino acid sequence within the third intracellular loop of the receptor.

[c] On the basis of sequence similarity, these receptor sequences have been identified as species homologs.

• These receptors belong to a discrete subfamily, exemplified by the metabotropic glutamate receptor mGluR1.

• • These receptors belong to a discrete subfamily, exemplified by the secretin receptor.

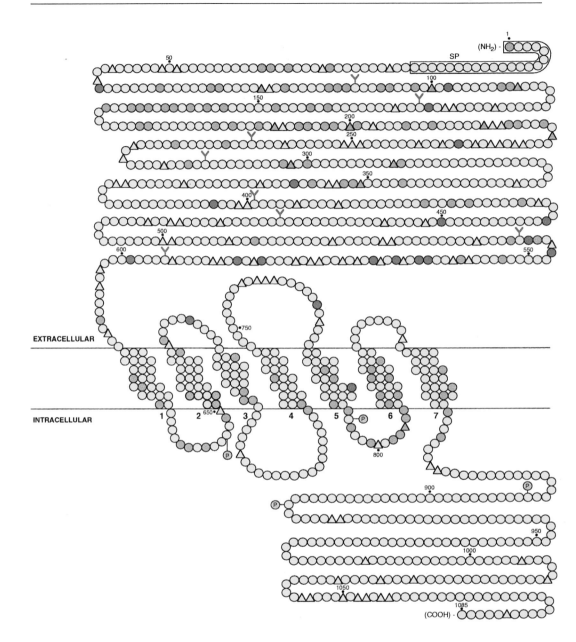

*Fig. 1.6. Model of the bovine Ca²⁺-sensing receptor. Putative transmembrane domains are defined and numbered consecutively and the signal peptide (SP) is boxed. Amino acid residues conserved among metabotropic glutamate receptors 1-6 and the bovine Ca²⁺-sensing receptor are depicted as blue circles, except for conserved acidic residues which are shown as blue triangles, and conserved Cys residues which are shown as green circles. Acidic residues occurring within the sequence of the bovine Ca²⁺-sensing receptor which are not conserved within the metabotropic glutamate receptor subfamily are shown as yellow triangles. Potential sites for N-linked glycosylation (Y) and for phosphorylation (P) are identified. [Reprinted with permission from Nature Volume 366, Brown EM, Gamba G, Riccardi D et al, Cloning and characterization of an extracellular Ca²⁺-sensing receptor from bovine parathyroid, Pages 575-580, Copyright (1993) Macmillan Magazines Limited].*

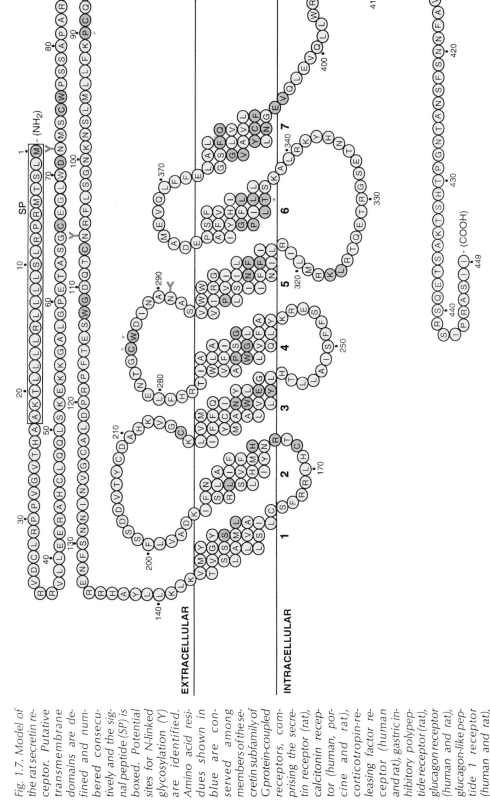

Fig. 1.7. Model of the rat secretin receptor. Putative transmembrane domains are defined and numbered consecutively and the signal peptide (SP) is boxed. Potential sites for N-linked glycosylation (Y) are identified. Amino acid residues shown in blue are conserved among members of the secretin subfamily of G protein-coupled receptors, comprising the secretin receptor (rat), calcitonin receptor (human, porcine and rat), corticotropin-releasing factor receptor (human and rat), gastric inhibitory polypeptide receptor (rat), glucagon receptor (human and rat), glucagon-like peptide 1 receptor (human and rat), growth hormone releasing hormone receptor (human, murine, porcine and rat), parathyroid hormone/parathyroid hormone-related peptide receptor (human, opossum and rat), pituitary adenylyl cyclase activating peptide receptor (rat) and vasoactive intestinal peptide receptor (human and rat). The diuretic hormone of tobacco hornworm (Manduca sexta) also belongs to this receptor subfamily and shows conservation of residues indicated in blue, except for residues marked with an asterisk.

A phylogenetic tree, generated by nucleotide sequence comparison of the visual pigment, adrenergic and muscarinic acetylcholine receptor genes, suggests that the opsins diverged from the catecholamine receptors between 100 and 150 million years ago.[263] The isolation of a *Dictyostelium discoideum* chemoattractant (cAMP) receptor, which bears structural and amino acid homology with receptors such as the $\beta_2$-adrenergic receptor, also suggests the age of the G protein-coupled receptor superfamily to be greater than 100 million years.[264] However, the $\alpha$ and a mating factor receptors of the yeast *Saccharomyces cerevisiae*, which also exhibit 7 TM topology, show no significant amino acid homology with other members of the superfamily.[265-267] The existence of these receptors clearly reflects the stability of this particular membrane organization throughout evolution but, as with members of the metabotropic glutamate and secretin receptor subfamilies, their evolutionary relationship to other members of the G protein-coupled receptor superfamily remains to be determined.

### i. Gene structure

The identification of receptors encoded by both intron-containing and intronless genes indicates a role for both gene duplication[268] and retroposition[269] mechanisms in the generation of diversity amongst members of the G protein-coupled receptor superfamily.

Along with few other mammalian genes, such as the genes encoding the histone proteins,[270] IFN-$\alpha$ and-$\beta$[271,272] and type X collagen,[273] the majority of G protein-coupled receptors are encoded by genes which do not contain introns within the coding region. Receptors encoded by genes which do contain introns include the vertebrate and invertebrate visual color pigments and several neuropeptide and peptide hormone receptors (see Fig. 1.8).[140,141,199,274-279] In addition, receptors with the same ligand binding specificity may be encoded by both intron-containing and intronless genes. For example, genes encoding the $\alpha_{1B}$-, $\alpha_{1C}$- and $\beta_3$-adrenergic,[190,280,281] dopamine $D_2$, $D_3$ and $D_4$[243] and serotonin 5-HT$_{2A}$, 5-HT$_{5A}$, 5-HT$_{5B}$,

5-HT$_6$ and 5-HT$_7$[282-285] receptor subtypes contain introns, while the $\alpha_{2A}$-, $\alpha_{2B}$-, $\beta_1$- and $\beta_2$-adrenergic,[190] dopamine $D_1$ and $D_5$[243] and serotonin 5-HT$_{1A}$[286] receptors are encoded by intronless genes (see Fig. 1.8). Conservation of sequence features characteristic of members of the superfamily (see Fig. 1.5) between receptors and receptor subtypes encoded by either intron-containing or intronless genes suggests evolution from a common ancestral gene. One or more of the intronless genes encoding G protein-coupled receptors may have evolved from a functionally related intron-containing gene through retroposition, involving reverse transcription of mRNA and incorporation into the genome, with subsequent gene duplication increasing the number of intronless genes. Consistent with this is the observation that the $\beta_2$-adrenergic receptor gene in both hamster and human is flanked by short direct repeat sequences, analogous to repeat sequences often found bordering processed genes.[287]

The odorant receptors in humans and rodents comprise a very large family of sequences. Encoded by an estimated 500 to 1000 genes, and incorporating pseudogenes, these receptor sequences may represent the largest known multigene family.[231,232,253,288] As with the majority of genes encoding G protein-coupled receptors, the coding regions of odorant receptor genes appear to be intronless. However, one or more introns may occur in the 5'-untranslated region of these genes. It has been suggested that gene rearrangement or gene conversion events, involving movement of a particular gene into an "active" region of the chromosome, or movement of control elements to the proximity of the gene to be expressed, could be responsible for selective gene expression among members of this receptor family.[288]

The occurrence of one or more introns within the 5'-untranslated region has been reported for a number of other G protein-coupled receptor genes whose coding region is intronless, such as the human and rodent angiotensin II AT$_1$,[289,290] murine angiotensin II AT$_2$,[291] human bradykinin

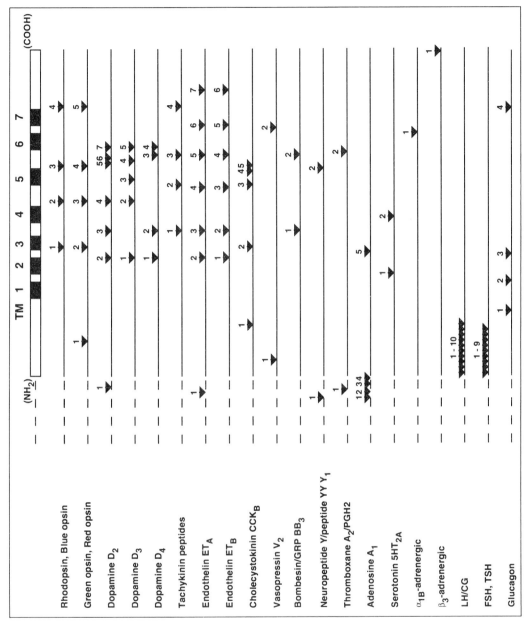

*Fig. 1.8. Intron/exon splice junctions for G protein-coupled receptor genes with splice junctions in the coding region. Intron/exon splice junctions are aligned with a schematic representation of a G protein-coupled receptor cDNA where the locations of transmembrane (TM) domains are shown as black boxes and other parts of the coding regions are shown as white boxes. N- and C-termini of the receptor are identified. The dashed line at the N-terminus of the receptor cDNA indicates 5'-untranslated cDNA sequence. The location of intron/exon splice sites is denoted with arrowheads and is compared among genes encoding rhodopsin and the visual color pigments,[460] dopamine,[319] tachykinin peptide (substance P, neurokinin A and neurokinin B),[461-463] endothelin $ET_A$ and $ET_B$,[275,276] cholecystokinin/gastrin $CCK_B$,[274] vasopressin $V_2$,[464] bombesin/gastrin-releasing peptide $BB_3$,[448] neuropeptide Y/peptide YY $Y_1$,[277,278] thromboxane $A_2$/prostaglandin (PG) $H_2$,[298] adenosine $A_1$,[296] serotonin $5HT_{2A}$,[282,283] $\alpha_{1B}$ and $\beta_3$-adrenergic,[280,281,465-468] luteinizing hormone/chorionic gonadotropin (LH/CG),[121] follicle-stimulating hormone (FSH),[121] thyroid-stimulating hormone (TSH),[121] and glucagon receptors.[170]*

B$_2$,[292] porcine muscarinic acetylcholine m2,[293] human N-formyl peptide[294] and human PAF[295] receptor genes, as well as for the human adenosine A$_1$, rat endothelin ET$_B$, human thromboxane A$_2$/PGH$_2$ and human and murine NPY/PYY Y$_1$ receptor genes, which contain at least one intron within the coding region (Fig. 1.8).[277,278,296-298] In the case of the human adenosine A$_1$, angiotensin II AT$_1$, porcine muscarinic acetylcholine m2, rat endothelin ET$_B$ and human PAF receptors, transcripts generated by alternative RNA processing within the 5'-untranslated region exhibit either differential abundance or differential tissue expression, consistent with a role for these sequences in the regulation of receptor levels.

Among G protein-coupled receptor genes which contain introns within the coding sequence there is considerable conservation of intron/exon splice junctions between receptors which exhibit the same or similar ligand specificity. This is seen clearly in the genes encoding the color visual pigments, the dopamine D$_2$, D$_3$ and D$_4$ receptor subtypes and the receptors for the tachykinin peptides substance P, NKA (substance K) and NKB (neuromedin K) (Fig. 1.8), and is consistent with evolution of these receptor families through gene duplication. There is also a significant degree of conservation of intron/exon splice junctions for G protein-coupled receptors as diverse as the visual color pigments, receptors for neurotransmitters such as dopamine and receptors for endothelin (ET$_A$ and ET$_B$) and the tachykinin peptides (Fig. 1.8), leading to the suggestion that a common precursor may have given rise to all of these receptors.[244,276]

A distinct subgroup of receptor genes may be represented by the $\alpha_{1B}$-adrenergic receptor (Fig. 1.8), with the single intron/exon splice site within the coding region conserved in $\alpha_{1B}$- and $\alpha_{1C}$-adrenergic receptor genes.[190,280] This gene structure is not observed in other members of the G protein-coupled receptors superfamily described to date, nor is it a feature of the $\alpha_{2A}$-, $\alpha_{2B}$-, $\beta_1$- or $\beta_2$-adrenergic receptor genes, which are intronless. The $\beta_3$-adrenergic receptor gene also contains a single intron/exon splice site in a novel location within the coding region.[281] However, the high degree of homology between $\beta_1$-, $\beta_2$- and $\beta_3$- adrenergic receptors[122] suggests evolution of the $\beta_3$-adrenergic receptor gene from an intronless precursor common to all three $\beta$-adrenergic receptors, with intron acquisition occurring in the $\beta_3$-adrenergic receptor gene at some time after duplication of the precursor gene.

Genes encoding receptors for the glycoprotein hormones LH/CG, FSH and TSH exhibit a similar structural organization (Fig. 1.8). The first 10 exons of the LH/CG receptor and the first 9 exons of the FSH and TSH receptors are functionally equivalent and encode the N-terminal extracellular domain of these receptors. The last exon encodes the remainder of all three receptors. Each of the exons contributing to the N-terminal domain of these receptors codes for monomers or multimers of the leucine-rich repeats present in this domain.[121] Since the N-terminal domain of these receptors incorporates the high-affinity ligand binding site, these receptor genes may have arisen by the addition of a coding sequence for a soluble, leucine-rich ligand-binding protein to an intronless gene coding for a product comprising TM segments with intervening cytoplasmic and extracellular domains that couples to G proteins.[279]

Of genes encoding receptors belonging to the secretin receptor subfamily, the intron/exon structure of the rat glucagon receptor gene has been elucidated (Fig. 1.8).[170] Splice variants have been described for other members of this receptor subfamily, including the rat calcitonin C1,[166] human CRF,[167] rodent GHRH[172,173] and rat PACAP type-I (PACAPR-2),[176] as well as for the rat metabotropic glutamate receptor mGluR1, indicating the presence of introns within the coding sequence of these receptor genes.[126,127]

## ii. Alternative splicing and receptor subtype diversity

G protein-coupled receptors comprise a number of structural and functional domains, including a ligand-binding site, TM segments and domains involved in interaction with G proteins and receptor desensitization (see Fig. 1.4). It is apparent from Figure 1.8 that intron/exon boundaries occur at or near putative TM boundaries or within discrete domains of the receptor such as the N-terminal extracellular domain, the third cytoplasmic loop or the C-terminal tail. Intron/exon organization may serve to arrange the receptor proteins into discrete functional domains, and it is becoming increasingly clear that alternative splicing contributes to the generation of receptor subtype diversity for many members of the receptor superfamily.

Splice variants have been described for a number of G protein-coupled receptors, including receptors for dopamine ($D_2$ and $D_3$), PGE$_2$ (EP$_3$), CCK/gastrin (CCK$_B$), somatostatin (SRIF$_{1A}$), substance P (NK$_1$), TRH, monocyte chemoattractant protein (MCP)-1, the glycoprotein hormones FSH, LH/CG and TSH, the calcitonin (C1), CRF, GHRH and PACAP type-I (PACAPR-2) receptors which belong the secretin receptor subfamily, and the metabotropic glutamate receptor mGluR1 (see Table 1.1).

The two forms of the D2 dopamine receptor, $D_{2A}$ (long) and $D_{2B}$ (short), differ by the presence of a 29 amino acid insertion in the third intracellular loop and arise from alternative splicing of a separate 87 bp exon (see Fig. 1.8).[243] The significance of this alternative splice site is highlighted by its preservation through evolution, from rodents to man,[299] and the demonstration of a functional difference in the efficiency of coupling to inhibition of adenylyl cyclase activity between these two receptor isoforms.[248,249] Splice variants of the somatostatin receptor SRIF$_{1A}$, which differ in length and sequence of the C-terminal domain, also exhibit differences in coupling efficiency, as well as in agonist-induced receptor desensitization, when heterologously expressed in CHO-K1 cells.[300] Splice variants of the bovine and murine PGE$_2$ EP$_3$ receptor, also differing in the C-terminal segment of the receptor, exhibit either differential G protein coupling efficiency[301]

or specificity.[250,251] Differential activation of intracellular effector systems has also been reported for splice variants of human substance P NK$_1$ receptor[302], which differ in the C-terminal domain. However, the significance of the existence of splice variants remains to be defined for some receptors. For example, splice variants of the TRH receptor, which differ in both length and sequence in the C-terminal domain, have indistinguishable electrophysiological properties.[303] Similarly, splice variants of the human MCP-1 receptor, which also differ in the length and sequence of the C-terminal domain, exhibit comparable activation properties,[304] while the functional properties of splice variants of the rabbit PGE$_2$ EP$_3$ receptor, which differ in the length and sequence of the C-terminal domain,[305] and of the human CCK$_B$ receptor, which differ by virtue of a pentapeptide insertion within the third intracellular loop,[274] have yet to be examined.

Alternative splicing gives rise to multiple forms of the dopamine $D_3$ receptor. Novel isoforms detected in rodents include receptors with either the deletion of 21 amino acids in the third intracellular loop, giving rise to $D_3$ (long) and $D_3$ (short) forms, or a 28 amino acid insertion in the first extracellular loop.[241,306] Transcripts encoding truncated receptor proteins which extend through only the first two or three TM domains of the native receptor or have a deletion extending from the second extracellular loop into TM5 have also been reported in rodents and man.[307-309] It remains to be established whether the production of these various transcripts is of biological significance, perhaps serving a regulatory role in controlling the abundance of native $D_3$ receptor transcripts, or in encoding shorter variants of the receptor which may play a role in cellular signaling events.

Expression of truncated receptor proteins has also been observed for FSH and LH/CG receptors. Splice variants of the FSH receptor comprise mature proteins representing 34% or 61% of the N-terminal extracellular domain of the native receptor,[310] while a splice variant of the LH/CG receptor is a secreted LH-binding protein encompassing the entire N-terminal extracellular domain of the native receptor.[311,312] The production of multiple mRNA transcripts and of splice variants that differ from native receptors by the deletion of exons encoding leucine-rich repeats within the N-terminal ligand binding domain has also been reported for FSH,[313,314] LH/CG[141,315] and TSH receptors.[140] Smaller receptor isoforms which possess moderate or high affinity binding sites are of potential physiological significance in hormonal regulation and modulation of cell responsiveness.[310]

Within the secretin subfamily of G protein-coupled receptors, differences in ligand binding or signaling properties have been demonstrated for a number of receptor subtypes which are generated by alternative splicing. For example, C1a and C1b splice variants of the rat calcitonin receptor, which differ by a 37 amino acid insertion in the second extracellular domain, have different affinity for calcitonin analogs.[166] The description of these receptor subtypes clearly identifies a functional domain of the calcitonin receptor which plays a major role in ligand recognition. The human CRF receptor also exists in two forms, differing by the insertion of 29 amino acids in the first extracellular loop, but whether or not functional differences exist between these receptor subtypes remains to be determined.[167] A truncated form of the rat CRF receptor, carrying deletion of sequence extending from the first extracellular loop to within TM4 and extending for only 39 amino acid residues beyond the splice junction, has also been reported.[169] Differential signal transduction capabilities have been reported for five splice variants of the rat PACAP type-I (PACAPR-2) receptor which differ in the length of the third intracellular loop.[176] The characterization of these receptor subtypes has identified a 28 amino acid sequence responsible for the impairment of coupling of the PACAP type-I (PACAPR-2) receptor to the stimulation of both adenylyl cyclase activity and PI metabolism. Splice variants of rodent GHRH receptors, which differ by the insertion of 41 amino acids in the third intracellular loop or by insertions in the first extracellular loop have been described, but their functional significance has not yet been determined.[172,173]

Splice variants mGluR1a, b and c differ in the length and sequence of the C-terminal tail.[126,127] Differences in signaling properties have been reported for mGluR1a and c subtypes.[127] It remains to be established if the mGluR1 receptor subtypes are differentially expressed in the central nervous system (CNS) The availability of sequence information provides the means for the generation of specific antipeptide antibodies which can be applied to address this question.

### iii. Polymorphic receptor variants

Two distinct types of DNA polymorphism within the dopamine $D_4$ receptor gene have been described. The first of these occurs as a 12 bp repeat, which codes for a sequence of four amino acids in the N-terminal extracellular segment of the receptor, bordering TM1.[316] The second type of polymorphic variation gives rise to a number of structural variants and is attributable to a 48 bp repeat sequence encoding 16 amino acid residues within the coding sequence encompassing the third intracellular loop of the dopamine $D_4$ receptor. Transcripts encoding 2, 4 and 7 copies of the repeat sequence have been isolated, and alleles containing 3, 5, 8 and 10 copies of the repeat sequence occur in genomic DNA of different individuals. The repeat sequences differ also in the sequence of the repeats and in the order in which they appear.[317,318] The functional significance of these receptor variants which differ in the third cytoplasmic loop is unclear, but may be of relevance in neuropsychiatric disorders. The structural variants appear to have altered ligand recognition properties and exhibit differential affinity for the dopamine receptor antagonist and antipsychotic therapeutic clozapine. Effects on intracellular signaling capabilities would also be expected with alteration of the size and sequence of the third intracellular loop.[319]

The existence of a variant of the dopamine $D_2$ receptor, which is characterized by a single amino acid substitution within the third intracellular loop, has also been reported.[242] The functional properties of this variant have not yet been determined. Genetic linkage analyses directed towards establishing the involvement of dopamine receptor subtypes and molecular variants in neuropsychiatric disorders will be addressed in chapter 4.

Two naturally occurring but uncommon variants of the human $\beta_2$-adrenergic receptor have been described, which exhibit a single amino acid change within either TM1 or TM4. No obvious phenotypic manifestations were reported to be associated with either receptor variant. However, the variant which carries an alteration within TM4 lacks the capability for high affinity ligand binding and is significantly impaired in functional coupling and agonist-stimulated receptor sequestration events. These deficits are believed to reflect the occurrence of this mutation within the putative ligand binding pocket of the receptor (see chapter 3),[320] highlighting the critical importance of this region of the receptor for normal receptor function.

## iv. Chromosomal localization

Considerable information regarding evolutionary relationships between various receptor genes in the context of evolution of the mammalian genome has been obtained from chromosomal localization studies in humans and comparative mapping in the murine genome. Chromosomal localization studies also provide the necessary framework for chromosomal mapping, allowing for correlation of genetic linkage analyses identifying specific disease loci with discrete molecular defects.

The chromosomal localization of genes encoding a number of G protein-coupled receptors is given in Table 1.2. The clustering of genes encoding receptors which are related in sequence and ligand specificity is evident in the case of $\alpha_{1B}$- and

$\beta_2$-adrenergic receptors, which occur spanning chromosomal position $5_{q23-q34}$, $\alpha_{2A}$- and $\beta_1$-adrenergic receptors at $10_{q24-q26}$, IL-8 A and B receptors at $2_{q35}$, and receptors for the chemoattractants C5a and N-formyl peptide at $19_{q13.3}$. Adrenergic $\alpha_{1B}$ and $\beta_2$ receptor genes are located within 300 kb of one another, while the distance between $\alpha_{2A}$- and $\beta_1$-adrenergic receptor genes is <225 kb.[321,322]

Close proximity of structurally and functionally related genes on the same chromosome suggests a strong evolutionary relationship, consistent with gene duplication events playing a major role in the generation of discrete molecular subtypes for adrenergic and other receptors. In addition, the high degree of sequence homology between adrenergic receptor subtypes whose genes are located on different chromosomes (see Table 1.2)[122] indicates the occurrence of chromosomal duplication during receptor subtype divergence. The $\alpha_{2A}$- and $\beta_1$-adrenergic receptor genes which are located on human chromosome 10 occur on mouse chromosome 19 and are believed to have arisen from duplication of an ancestral precursor gene, with $\alpha_{1B}$- and $\beta_2$-adrenergic receptor genes, located on human chromosome 5, arising subsequently as a consequence of chromosomal duplication. The $\alpha_{1B}$- and $\beta_2$-adrenergic receptor genes are located on murine chromosomes 11 and 19, respectively, but the organization of genes linked to these loci on the murine chromosomes resembles the organization of homologous genes on human chromosome 5. Thus, while chromosomal rearrangement has occurred in the mammalian lineage leading to the mouse, linkage groups have been conserved during evolution of human and murine genomes.[321,323,324]

In contrast to chromosomal co-localization for adrenergic receptor subtypes and structurally related IL-8 and chemoattractant receptors, the genes encoding many G protein-coupled receptors which are structurally and functionally related are located on different human chromosomes. These include the endothelin $ET_A$ and $ET_B$, and somatostatin receptor subtypes, as well as receptors for the tachykinin peptides substance P ($NK_1$) and NKA (substance K; $NK_2$), and $D_1$-$D_5$ receptor subtypes for the neurotransmitter dopamine. Sequence similarities suggest the involvement of gene duplication events in the generation of the receptor families, with chromosomal rearrangement accounting for the dispersion of individual receptor genes throughout the genome.

The cluster of genes encoding IL-8 A and B receptors contains an inactivated pseudogene, *IL8RAP*, which is structurally related to the IL-8 B receptor. Two pseudogenes of the dopamine $D_5$ receptor, *DRD5P1* and *DRD5P2*, have also been

## Table 1.2. Chromosomal localization of G protein-coupled receptor gene loci

| | | | Chromosome | Reference* |
|---|---|---|---|---|
| **PEPTIDES AND PEPTIDE HORMONES** | | | | |
| Angiotensin II | $AT_1$ | AGTR1 | 3 q21-q25 | |
| | mas | MAS1 | 6 q24-q27 | 447 |
| Bombesin/GRP | $BB_2$ | GRPR | X p11-q11 | 447 |
| | $BB_3$ | BRS3 | X q26-q28 | 448 |
| Bombesin/neuromedin B | $BB_1$ | NMBR | 6 q21-qter | |
| Bradykinin | $B_2$ | BDKRB2 | 14 | |
| CCK/gastrin | $CCK_B$ | CCKBR | 11 p15.4 | |
| Complement C5a | | C5R1 | 19 q13.3 | 449 |
| Endothelin | $ET_A$ | EDNRA | 4 | |
| | $ET_B$ | EDNRB | 13 | |
| FSH | | FSHR | 2 p21-p16 | |
| GLP 1 | | GLP1R | 6 p21.1 | 450 |
| IL-8 | A | IL8RA | 2 q35 | |
| | B[a] | IL8RB | 2 q35 | |
| | | IL8RAP | 2 q35 | 451 |
| LH/CG | | LHCGR | 2 p21 | |
| MIP-1α/RANTES | | CMKBR1 | 3 p21 | 452 |
| MSH | MC-2 | MC2R | 18 p11.2-pter | 453 |
| | MC-4 | MC4R | 18 q21.3 | 151 |
| N-formyl peptide[b] | | FPR1 | 19 q13.3 | 449 |
| | | FPRH1 | 19 q13.3 | 449 |
| | | FPRH2 | 19 q13.3 | 449 |
| NKA (substance K) | $NK_2$ | TAC2R | 10 q11-q21 | |
| NPY | $Y_1$ | NPY1R | 4 q31.3-q32 | |
| Opioid | δ | | 1 p34.3-p36.1 | 454,455 |
| | κ | | 8 | 454 |
| | μ | | 6 q25 | 454 |
| PTH/PTHrP | | PTHR | 3 p21.1-p22 | 376 |
| Somatostatin | $SRIF_{1A}$ | SSTR2 | 17 q24 | |
| | $SRIF_{1B}$ | SSTR5 | 20 p11.2 | |
| | $SRIF_{1C}$ | SSTR3 | 22 q13.1 | |
| | $SRIF_{2A}$ | SSTR1 | 14 q13 | |
| | $SRIF_{2B}$ | SSTR4 | 20 | |
| Substance P | $NK_1$ | TAC1R | 2 | |
| Thrombin | | F2R | 5 q13 | |
| TSH | | TSHR | 14 q31 | |
| Vasopressin/oxytocin | $V_2$ | AVPR2 | X q28 | |
| **NEUROTRANSMITTERS** | | | | |
| Adenosine | A1 | ADORA1 | 22 q11.2-q13.1 | |
| | $A2_A$ | ADORA2A | 11 q11-q13 | |
| | $A2_B$ | ADORA2L | 10 q25.3-q26.3 | |

## Table 1.2. continued

| | | | | | Reference* |
|---|---|---|---|---|---|
| Adrenergic | $\alpha_{1B}$ | ADRA1B | 5 | q32-q34 | 321 |
| | $\alpha_{1C}$ | ADRA1C | 8 | p21-p11.2 | |
| | $\alpha_{1D}$ | ADRA1D | 20 | p13 | 322 |
| | $\alpha_{2A}$ | ADRA2A | 10 | q24-q26 | |
| | $\alpha_{2B}$ | ADRA2B | 2 | p13-q13 | |
| | $\alpha_{2C}$ | ADRA2C | 4 | p16.3-p15 | |
| | $\beta_1$ | ADRB1 | 10 | q24-q26 | |
| | $\beta_2$ | ADRB2 | 5 | q31-q32 | |
| | $\beta_3$ | ADRB3 | 8 | p12-p11 | 456 |
| Dopamine | $D_1$ | DRD1 | 5 | q35.1 | 457 |
| | $D_2$ | DRD2 | 11 | q23.1 | |
| | $D_3$ | DRD3 | 3 | q13.3 | |
| | $D_4$ | DRD4 | 11 | p15.5 | |
| | $D_5$[a] | DRD5 | 4 | p15.3-p15.1 | |
| | | DRD5P1 | 2 | p11.2-p11.1 | |
| | | DRD5P2 | 1 | q21.1 | |
| Muscarinic acetylcholine | m1 | CHRM1 | 11 | q12-q13 | |
| | m2 | CHRM2 | 7 | q35-qter | |
| | m3 | CHRM3 | 1 | q41-q44 | |
| | m4 | CHRM4 | 11 | p12-p11.2 | |
| | m5 | CHRM5 | 15 | q26 | |
| Serotonin | $5\text{-}HT_{1A}$ | HTR1A | 5 | qcen-q11 | |
| | $5\text{-}HT_{1D\alpha}$[a] | HTR1D | 1 | p36.3-p34.3 | |
| | | HTR1DP1 | 12 | | |
| | $5\text{-}HT_{1D\beta}$ | HTR1B | 6 | q13 | |
| | $5\text{-}HT_{2A}$ | HTR2 | 13 | q14-q21 | |
| | $5\text{-}HT_{2C}$ | HTR1C | X | q24 | |
| **INORGANIC IONS** | | | | | |
| Ca$^{2+}$ | | | 3 | | 410 |
| **LIPID MEDIATORS** | | | | | |
| Anandamide | | CNR1 | 6 | q14-q15 | |
| PAF | | PTAFR | 1 | | |
| Thromboxane A2/PGH$_2$ | | TBXA2 | 19 | p13.3 | 298,458 |
| **SENSORY STIMULI** | | | | | |
| Odorant | | OLFR1 | 17 | p13-p12 | |
| Rhodopsin | | RHO | 3 | q21-q24 | |
| Blue opsin (Blue cone pigment) | | BCP | 7 | q31.3-q32 | 459 |
| Green opsin (Green cone pigment) | | GCP | X | q28 | |
| Red opsin (Red cone pigment) | | RCP | X | q28 | |
| **ORPHAN RECEPTORS** | | | | | |
| RDC1 | | D2S87E | 2 | q37 | |
| D2S201E | | D2S201E | 2 | q21 | |

Receptors are categorized according to endogenous ligand. Receptor subtype designations are based primarily on the nomenclature tabulation of Watson and Girdlestone (1994).[230] Gene loci are as listed in the Genome Data Base (GDB™).

* Citations for chromosomal localization of gene loci are available in GDB entries and only literature references in addition to those available from this source are listed in this Table.

[a] Denotes gene encoding receptor for which one or more pseudogenes have been identified.

[b] Denotes gene encoding receptor for which structurally related orphan receptor homologs have been identified.

nine $D_5$ receptor different chromo- :ional homolog (see ctional receptor gene s are surrounded by mic sequence, indica- lutionary origin. The pseud~ to be specific to humans. They do not encode functional receptors, but are transcribed in a number of brain areas and may play a role in regulating the level of expression of the functional $D_5$ receptor.[319]

Correlations between the chromosomal localization of G protein-coupled receptor genes and genetic linkage studies aimed at identifying particular disease loci have been made for a number of receptors and pathological conditions. For example, molecular cloning of the vasopressin $V_2$ receptor, which enabled localization of the vasopressin $V_2$ receptor gene to the long arm of the X chromosome (see Table 1.2),[325] identified a potential link to familial vasopressin-resistant nephrogenic diabetes insipidus (NDI). A number of mutations in this receptor which co-segregate with the disease have now been reported. Molecular defects associated with this and other G protein-coupled receptors, as well as therapeutic potential derived from the molecular characterization of receptor subtypes, will be discussed in detail in chapter 4.

## 4. PHYSIOLOGICAL SIGNIFICANCE OF MOLECULAR SUBTYPES

The molecular characterization of receptors and receptor subtypes clearly indicates an apparent redundancy of subtypes for a large number of G protein-coupled receptors (see Sections 1 and 3, above; see Table 1.1). However, an appreciation of the physiological significance of receptor subtype diversity is beginning to be attained from determination of the functional capabilities of particular receptor subtypes and the regulation of their expression in a tissue- or developmental-specific manner.

For a number of G protein-coupled receptors activated by the same ligand,

different receptor subtypes have opposite effects on the same intracellular signaling pathway or modulate different intracellular signaling pathways. This provides a means for a diversity of responses to a single ligand, depending upon expression of particular receptor subtypes within individual cell types or tissues. For example, the existence of dopamine $D_1$-like ($D_1$ and $D_5$) and $D_2$-like ($D_2$ and $D_4$) receptors allows dopamine to either stimulate or inhibit adenylyl cyclase activity in the cell.[230] Similarly, the interaction of adenosine with $A_{2A}$ or $A_{2B}$ receptors results in stimulation of adenylyl cyclase activity, while activation of adenosine $A_1$ receptors results in inhibition of adenylyl cyclase activity.[155] An example of coupling to different intracellular signaling pathways is seen with the muscarinic acetylcholine receptors, where m1, m3 and m5 muscarinic acetylcholine receptors couple to stimulation of PI metabolism, and m2 and m4 receptors couple to inhibition of adenylyl cyclase activity.[326] Additional complexity exists with receptors of a particular subtype being capable of activating multiple signal transduction pathways. For example, activation of the dopamine $D_2$ receptor subtypes may result in stimulation of $K^+$ channel activity, inhibition of $Ca^{2+}$ channel activity and stimulation of arachidonic acid metabolism, in addition to inhibition of adenylyl cyclase activity.[230,327] A similar situation exists for the adenosine $A_1$ receptor, which couples to inhibition of adenylyl cyclase activity, as well as stimulation of both $K^+$ channel activity and PI metabolism.[155,328]

Additional properties of receptor subtypes which have substantial impact upon their functional capabilities in a physiological context include their inherent responsiveness to ligand availability and the duration of the response elicited. For a number of receptors, differences exist in the affinity of receptor subtypes for their endogenous ligand. For example, the $D_1$-like dopamine receptor subtypes $D_1$ and $D_5$ exhibit a 10-fold difference in affinity for dopamine, while up to 150-fold differences in affinity for dopamine are observed between the

$D_2$-like receptor subtypes $D_2$, $D_3$ and $D_4$.[243] Similarly, the adenosine $A_{2A}$ and $A_{2B}$ receptor subtypes represent high and low affinity forms, respectively.[131,132] Such receptor subtype diversity provides the means for receptor activation at low or high concentrations of ligand. Receptor responsiveness is regulated temporally by the process of desensitization, whereby receptors become refractory to further stimulation in the continued presence of activating ligand. This occurs as a consequence of phosphorylation at specific sites within the receptor sequence which are recognized by protein kinases such as β-ARK, PKA or PKC (see Fig. 1.5).[191] It has been clearly demonstrated for the $β_1$-, $β_2$- and $β_3$-adrenergic receptor subtypes that the rate and extent of desensitization depend upon the type and number of potential phosphorylation sites (see chapter 3).[188,190] Differences in potential regulatory sequences in the dopamine and other receptor subtypes are also believed to represent the means for differential regulation of desensitization mechanisms.[243,329]

The availability of specific sequence probes attained by molecular cloning of receptors has allowed investigation of sites of expression of discrete receptor subtypes. For a number of receptors, including adrenergic, dopamine, muscarinic acetylcholine and metabotropic glutamate receptors, different receptor subtypes with similar signal transduction properties exhibit differential tissue expression. For example, both $β_1$- and $β_2$-adrenergic receptor subtypes, which couple to stimulation of adenylyl cyclase activity, are expressed in human myocardial tissue, where they are thought to mediate increases in heart rate and myocardial inotropy. $β_2$-adrenergic receptors also occur in the lung and in peripheral vasculature, where they mediate bronchodilation and vasodilation, respectively.[190,330] Both m2 and m4 muscarinic acetylcholine receptor subtypes couple to inhibition of adenylyl cyclase activity. However, within the CNS, the m4 subtype is expressed in the cortex, striatum and hippocampus, while the m2 subtype occurs only in the

medial septum, pons and thalamus. The m2 subtype is the only muscarinic acetylcholine receptor subtype known to be expressed in cardiac atrial tissue, where its additional functional capability of stimulating the opening of $K^+$ channels is believed to play a role in the hyperpolarization process.[195,326] Similarly, metabotropic glutamate receptors with similar signal transduction properties and agonist selectivity are differentially expressed in the CNS.[331] The further characterization of functional properties of these receptor subtypes and correlation with their sites of expression should provide the means for elucidation of the precise physiological role of glutamate in CNS function.

Different regulatory signals within the genes encoding receptor subtypes allow for differential regulation and tissue-specific expression. For example, expression of the $β_3$-adrenergic receptor is highly tissue-specific, with the majority of $β_3$-adrenergic receptor mRNA being found in brown adipose tissue. In mammals other than rodents, brown adipose tissue occurs mainly in neonates or in rare pathological conditions such as pheochromocytoma. Murine 3T3-F44-2A fibroblasts normally express only $β_1$- and $β_2$-adrenergic receptor mRNA, but when induced to differentiate into adipocytes, $β_2$-adrenergic receptor mRNA levels become barely detectable and the cells express predominantly $β_3$-adrenergic receptors.[190] Steroid hormone-responsive elements and cAMP-responsive elements (CRE) have been identified in the 5'-upstream regulatory region of a number of receptor genes, allowing for regulation of receptor expression in response to changes in circulating hormone levels or intracellular levels of cAMP, respectively. For example, the $β_2$-adrenergic receptor gene contains several glucocorticoid-responsive elements (GRE) consensus sequences that may interact with the glucocorticoid receptor, while the $β_3$-adrenergic receptor gene contains three CRE which function in agonist-induced increased transcription of the receptor gene as a consequence of elevation of cAMP levels.[190]

Continuing isolation of discrete molecular subtypes of receptors and characterization of their sites of expression will undoubtedly also allow delineation of the role specific receptors may play in mediating complex physiological processes. For example, transcripts encoding members of the odorant receptor family, which comprises over 100 members closely related to one another in primary sequence (see Table 1.1),[122] have been detected in male germ cells. If these receptors are effectively translated and expressed on the cell surface of spermatozoa, they may interact with one or more specific ligands within the female genital tract and so mediate chemotaxis of spermatozoa during fertilization. The search for novel ligands may provide substantial new knowledge on the process of fertilization and mechanisms underlying human fertility.[253]

Receptors for a number of neurotransmitters, including the $\alpha_2$-adrenergic and dopamine receptor subtypes, are known to exist in both presynaptic and postsynaptic locations, where they mediate different functions.[188,243] Thus, presynaptic $\alpha_2$-adrenergic receptors inhibit neurotransmitter release and play a role in the modulation of sympathetic tone, while postsynaptic $\alpha_2$-adrenergic receptors mediate vasoconstriction by modulation of smooth muscle tone in arterioles.[188] In many instances, the occurrence of receptors at presynaptic or postsynaptic locations reflects differential tissue expression. However, for neurotransmitter receptors occurring within the CNS, presynaptic or postsynaptic location involves the targeting of receptor subtypes towards the dendrites and cell body or towards the nerve terminal, respectively. The mechanisms underlying such discrete intracellular localization remain to be defined.

## 5. ORPHAN RECEPTORS

Orphan receptors comprise novel receptor sequences whose assignment to the G protein-coupled receptor superfamily is currently based on sequence features alone, but whose activating ligand remains unidentified. Such receptors have been isolated mainly by application of low stringency hybridization and PCR amplification using degenerate oligonucleotide primers. Endogenous ligands have been identified only for relatively few receptors originally isolated as orphan sequences. Examples of such receptors are the clones originally isolated from canine thyroid tissue and designated RDC4, RDC7 and RDC8, which have been identified as serotonin 5-HT$_{1D}$, adenosine A$_{2A}$ and adenosine A$_1$ receptors, respectively,[227,332] as well as the human ACTH, MSH MC-1 and somatostatin SRIF$_{2A}$, and the rat adenosine A$_3$, cannabinoid (anadamide) and NPY/PYY Y$_1$ receptors.[332]

There are several orphan receptors which exhibit sequence similarity to receptors whose ligand is known. For example, a receptor originally identified as the NPY Y$_3$ receptor,[333] which has subsequently been shown to lack NPY binding specificity or responsiveness,[236,334] bears striking homology to the IL-8 receptor family and could represent another receptor which interacts with a member or members of the chemoattractant cytokine (chemokine) family of ligands.[335,336] The high level of expression of this orphan receptor, designated D2S201E (see Table 1.2), in cells of hemopoietic origin,[337-339] is consistent with this suggestion. Related to the human N-formyl peptide receptor are the putative receptors FPRL1, FPRH1 and FPRH2 (see Tables 1.1 and 1.2). FPRL1 is expressed on neutrophils, and its isolation may indicate the existence of a novel chemotactic factor involved in recruitment of neutrophils to sites of inflammation.[234] However, structural similarity does not always provide a ready identification of ligand specificity for orphan receptors. For example, two orphan receptors denoted "rat thoracic aorta" (RTA) and "*mas*-related gene" (*mrg*) are structurally most closely related to the receptor encoded by the *mas* oncogene, which has been reported to be an angiotensin II receptor.[340-342] The orphan receptors RTA and *mrg* do not bind angiotensin II and their endogenous ligand remains unknown.

The cloning of the cannabinoid receptor[224] and investigation of its functional significance allowed the subsequent identification of the eicosanoid anandamide as a bioactive molecule.[225,226] The availability of orphan receptor sequences and the ability to express these receptors in heterologous systems such as *Xenopus* oocytes and a variety of mammalian cell lines provide the basis for a continuing search for bioactive molecules in complex biological fluids and tissue extracts. Recently developed random peptide libraries[343-346] could also be implemented in screening procedures for potentially active novel ligands. Such avenues of investigation will continue to expand our definitions of endogenous ligands, potentially leading to the description of additional intercellular communication pathways.

## SUMMARY

The isolation and characterization of discrete members of the G protein-coupled receptor superfamily have been made possible by the application of a variety of molecular cloning approaches and considerable insight into both the structural and functional diversity of this receptor superfamily has been attained. Existing pharmacological definitions of subtypes of receptor for known endogenous ligands have been extended and functional analyses have demonstrated that the diversity of function elicited by individual ligands derives at least in part from the existence of distinct molecular subtypes of receptor. In addition, novel receptor sequences, whose activating ligand remains unidentified, have been isolated. The assignment of these "orphan" receptors to the G protein-coupled receptor superfamily is currently based on sequence features alone, but continuing investigations of their functional significance will undoubtedly extend current knowledge of both known and yet to be identified bioactive molecules mediating physiological responses.

## REFERENCES

1. Birnbaumer L, Abramowitz J, Brown AM. Receptor-effector coupling by G proteins. Biochim Biophys Acta 1990; 1031:163-224.

2. Williams AF. A year in the life of the immunoglobulin superfamily. Immunol Today 1987; 8:298-303.

3. Williams AF, Barclay AN. The immunoglobulin superfamily - domains for cell surface recognition. Annu Rev Immunol 1988; 6:381-405.

4. Buck CA. Immunoglobulin superfamily: structure, function and relationship to other receptor molecules. Sem Cell Biol 1992; 3:179-188.

5. Cunningham BA, Hemperly JJ, Murray BA et al. Neural cell adhesion molecule: structure, immunoglobulin-like domains, cell surface modulation, and alternative RNA splicing. Science 1987; 236:799-806.

6. Turner ML. Cell adhesion molecules: a unifying approach to topographic biology. Biol Rev 1992; 67:359-377.

7. Takeichi M. The cadherins: cell-cell adhesion molecules controlling animal morphogenesis. Development 1988; 102:639-655.

8. Takeichi M. Cadherins: a molecular family important in selective cell-cell adhesion. Annu Rev Biochem 1990; 59:237-258.

9. Takeichi M. Cadherin cell adhesion receptors as a morphogenetic regulator. Science 1991; 251:1451-1455.

10. Fleming S. Cellular functions of adhesion molecules. J Pathol 1990; 161:189-190.

11. Hynes RO. Specificity of cell adhesion in development: the cadherin superfamily. Curr Opin Genet Dev 1992; 2:621-624.

12. Kemler R. Classical cadherins. Sem Cell Biol 1992; 3:149-155.

13. Rosen SD. The LEC-CAMs: an emerging family of cell-cell adhesion receptors based upon carbohydrate recognition. Am J Respir Cell Mol Biol 1990; 3:397-402.

14. Rosen SD. L-selectin and its biological ligands. Histochemistry 1993; 100:185-191.

15. Springer TA, Lasky LA. Cell adhesion: sticky sugars for selectins. Nature 1991; 349:196-197.

16. Lasky LA. Selectins: interpreters of cell-specific carbohydrate information during inflammation. Science 1992; 258:964-969.

17. Vestweber D. Selectins: cell surface lectins which mediate the binding of leukocytes to endothelial cells. Sem Cell Biol 1992; 3:211-220.

18. Hynes RO. Integrins: a family of cell surface receptors. Cell 1987; 48:549-554.

19. Hemler ME. Adhesive protein receptors on hematopoietic cells. Immunol Today 1988; 9:109-113.

20. Albelda SM, Buck CA. Integrins and other cell adhesion molecules. FASEB J 1990; 4:2868-2880.

21. Werb Z, Tremble P, Damsky CH. Regulation of extracellular matrix degradation by cell-extracellular matrix interactions. Cell Differ Dev 1990; 32:299-306.

22. Bosman FT. Integrins: cell adhesives and modulators of cell function. Histochem J 1993; 25:469-477.

23. Haynes BF, Telen MJ, Hale LP et al. CD44—a molecule involved in leukocyte adherence and T-cell activation. Immunol Today 1989; 10:423-428.

24. Lewinsohn DM, Nagler A, Ginzton N et al. Haematopoietic progenitor cell expression of the H-CAM (CD44) homing-associated adhesion molecule. Blood 1990; 75:589-595.

25. Haynes BF, Liao H-X, Patton KL. The transmembrane hyaluronate receptor (CD44): multiple functions, multiple forms. Cancer Cells 1991; 3:347-350.

26. Gallagher JT. The protein and proteoglycan guises of Hermes/CD44. Glycobiology 1992; 2:93-94.

27. Underhill C. CD44: the hyaluronan receptor. J Cell Sci 1992; 103:293-298.

28. Brown MS, Goldstein JL. A receptor-mediated pathway for cholesterol homeostasis. Science 1986; 232:34-47.

29. Schneider WJ. The low density lipoprotein receptor. Biochim Biophys Acta 1989; 988:303-317.

30. Soutar AK, Knight BL. Structure and regulation of the LDL-receptor and its gene. Brit Med Bull 1990; 46:891-916.

31. McClelland A, Kühn LC, Ruddle FH. The human transferrin receptor gene: genomic organization and the complete primary structure of the receptor deduced from a cDNA sequence. Cell 1984; 39:267-274.

32. Aisen P. Entry of iron into cells: a new role for the transferrin receptor in modulating iron release from transferrin. Ann Neurol 1992; 32:S62-S68.

33. Testa U, Pelosi E, Peschle C. The transferrin receptor. Crit Rev Oncog 1993; 4:241-276.

34. Bell GI, Kayano T, Buse JB et al. Molecular biology of mammalian glucose transporters. Diabetes Care 1990; 13:198-208.

35. Thorens B, Charron MJ, Lodish HF. Molecular physiology of glucose transporters. Diabetes Care 1990; 12:209-218.

36. Baldwin SA. Molecular mechanisms of sugar transport across mammalian and microbial cell membranes. Biotechnol Applied Biochem 1990; 12:512-516.

37. Kong C-T, Yet S-F, Lever JE. Cloning and expression of a mammalian Na+/amino acid cotransporter with sequence similarity to Na+/glucose cotransporters. J Biol Chem 1993; 268:1509-1512.

38. Souba WW, Pacitti AJ. How amino acids get into cells: mechanisms, models, menus, and mediators. J Parenter Enter Nutrition 1992; !6:569-578.

39. Kim JW, Closs EI, Albritton LM et al. Transport of cationic amino acids by the mouse ecotropic retrovirus receptor. Nature 1991; 352:725-728.

40. Wang H, Kavanaugh MP, North RA et al. Cell-surface receptor for ecotropic murine retroviruses is a basic amino acid transporter. Nature 1991; 352:729-731.

41. Christensen HH. A retrovirus uses a cationic amino acid transporter as a cell surface receptor. Nutr Rev 1992; 50:47-48.

42. Tate SS, Yan N, Udenfriend S. Expression cloning of a Na+-independent neutral amino acid transporter from rat kidney. Proc Natl Acad Sci USA 1992; 89:1-5.

43. Wells RG, Hediger MA. Cloning of a rat kidney cDNA that stimulates dibasic and neutral amino acid transport and has sequence similarity to glucosidases. Proc Natl Acad Sci USA 1992; 89:5596-5600.

44. Bertran J, Werner A, Moore ML et al. Expression cloning of a cDNA from rabbit kidney cortex that induces a single transport system for cystine and dibasic and neutral amino acids. Proc Natl Acad Sci USA 1992; 89:5601-5605.

45. You G, Smith CP, Kanai Y et al. Cloning and characterization of the vasopressin-regulated urea transporter. Nature 1993;

365:844-847.

46. Schloss P, Mayser W, Betz H. Neurotransmitter transporters. A novel family of integral plasma membrane proteins. FEBS Lett 1992; 307:76-80.

47. Kanner BI. Glutamate transporters from brain. A novel neurotransmitter transporter family. FEBS Lett 1993; 325:95-99.

48. Amara S, Arriza JL. Neurotransmitter transporters: three distinct gene families. Curr Opin Neurobiol 1993; 3:337-344.

49. Hyde SC, Emsley P, Hartshorn MJ et al. Structural model of ATP-binding proteins associated with cystic fibrosis, multidrug resistance and bacterial transport. Nature 1990; 346:362-365.

50. Collins F. Cystic fibrosis: molecular biology and therapeutic implications. Science 1992; 256:774-779.

51. Kuchler K, Thorner J. Secretion of peptides and proteins lacking hydrophobic signal sequences: the role of adenosine triphosphate-driven membrane translocators. Endocr Rev 1992; 13:499-514.

52. Gillis S. Cytokine receptors. Curr Opin Immunol 1991; 3:315-319.

53. Olsson I, Gullberg U, Lantz M et al. The receptors for regulatory molecules of hematopoiesis. Eur J Haematol 1992; 48:1-9.

54. Miyajima A, Kitamura T, Harada N et al. Cytokine receptors and signal transduction. Annu Rev Immunol 1992; 10:295-331.

55. Hall AK, Rao MS. Cytokines and neurokines: related ligands and related receptors. Trends Neurosci 1992; 15:35-37.

56. Dower SK. Cytokine receptor families. Adv Second Messenger Phosphoprotein Res 1993; 28:19-25.

57. Taga T, Kishimoto T. Cytokine receptors and signal transduction. FASEB J 1993; 7:3387-3396.

58. Kaczmarski RS, Mufti GJ. The cytokine receptor superfamily. Blood Rev 1991; 5:193-203.

59. Stahl N, Yancopoulos GD. The alphas, betas, and kinases of cytokine receptor complexes. Cell 1993; 74:587-590.

60. Kitamura T, Ogorochi T, Miyajima A. Multimeric cytokine receptors. Trends Endocrinol Metab 1994; 5:8-14.

61. Yarden Y, Ullrich A. Growth factor receptor tyrosine kinases. Annu Rev Biochem 1988; 57:443-478.

62. Yarden Y, Ullrich A. Molecular analysis of signal transduction by growth factors. Biochemistry 1988; 27:3113-3119.

63. Cadena DL, Gill GN. Receptor tyrosine kinases. FASEB J 1992; 6:2332-2337.

64. Miki T, Felming TP, Bottaro DP et al. Expression cDNA cloning of the KGF receptor by creation of a transforming autocrine loop. Science 1991; 251:72-75.

65. Givol D, Yayon A. Complexity of FGF receptors: genetic basis for structural diversity and functional specificity. FASEB J 1992; 6:3361-3369.

66. Mathews LS, Vale WW. Expression cloning of an activin receptor, a predicted transmembrane serine kinase. Cell 1991; 65:973-982.

67. Attisano L, Wrana JL, Cheifetz S et al. Novel activin receptors: distinct genes and alternative mRNA splicing generate a repertoire of serine/threonine kinase receptors. Cell 1992; 68:97-108.

68. O'Grady P, Liu Q, Huang SS et al. Transforming growth factor β (TGF-β) type V receptor has a TGF-β-stimulated serine/threonine-specific autophosphorylation activity. J Biol Chem 1992; 267:21033-21037.

69. Hunter T. Protein-serine kinase receptors? Curr Biol 1991; 1:15-16.

70. Massagué J, Andres J, Attisano L et al. TGF-β receptors. Mol Reprod Dev 1992; 32:99-104.

71. Matsuzaki K, Xu J, Wang F et al. A widely expressed transmembrane serine/threonine kinase that does not bind activin, inhibin, transforming growth factor β, or bone morphogenic factor. J Biol Chem 1993; 268:12719-12723.

72. Ebner R, Chen R-H, Shum L et al. Cloning of a type I TGF-β receptor and its effect on TGF-β binding to the type II receptor. Science 1993; 260:1344-1348.

73. Janeway CA. The T cell receptor as a multicomponent signalling machine: CD4/CD8 coreceptors and CD45 in T cell activation. Annu Rev Immunol 1992; 10:645-674.

74. Trowbridge IS, Ostergaard HL, Johnson P. CD45: a leukocyte-specific member of the

protein tyrosine phosphatase family. Biochim Biophys Acta 1991; 1095:46-56.

75. Trowbridge IS, Johnson P, Ostergaard H et al. Structure and function of CD45: a leukocyte-specific protein tyrosine phosphatase. Adv Exp Med Biol 1992; 323:29-37.

76. Weaver CT, Pingel JT, Nelson JO et al. CD45: a transmembrane protein tyrosine phosphatase involved in the transduction of antigenic signals. Biochem Soc Trans 1992; 20:169-174.

77. Koretzky GA. Role of CD45 tyrosine phosphatase in signal transduction in the immune system. FASEB J 1993; 7:420-426.

78. Garbers DL. Guanylate cyclase, a cell surface receptor. J Biol Chem 1989; 264:9103-9106.

79. Garbers DL. Guanylate cyclase receptor family. Recent Prog Horm Res 1990; 46:85-97.

80. Garbers DL, Koesling D, Schultz G. Guanylyl cyclase receptors. Mol Biol Cell 1994; 5:1-5.

81. Levin ER. Natriuretic peptide C-receptor: more than a clearance receptor. Am J Physiol 1993; 264:E483-E489.

82. Strange PG. The structure and mechanism of neurotransmitter receptors. Implications for the structure and function of the central nervous system. Biochem J 1988; 249:309-318.

83. North RA. Neurotransmitters and their receptors: from the clone to the clinic. Sem Neurosci 1989; 1:81-90.

84. Betz H. Ligand-gated ion channels in the brain: the amino acid receptor superfamily. Neuron 1990; 5:383-392.

85. Dingledine R, Myers SJ, Nicholas RA. Molecular biology of mammalian amino acid receptors. FASEB J 1990; 4:2636-2645.

86. Seeburg PH. The molecular biology of mammalian glutamate receptor channels. Trends Neurosci 1993; 16:359-365.

87. Brisson A, Unwin PNT. Quarternary structure of the acetylcholine receptor. Nature 1985; 315:474-477.

88. Pritchett DB, Sontheimer H, Shivers BD et al. Importance of a novel GABA$_A$ receptor subunit for benzodiazepine pharmacology. Nature 1989; 338:582-585.

89. Lüddens H, Pritchett DB, Köhler M et al. Cerebellar GABA$_A$ receptor selective for a behavioural alcohol antagonist. Nature 1990; 346:648-651.

90. Unwin N. Neurotransmitter action: opening of ligand-gated ion channels. Cell 1993; 72 Suppl:31-41.

91. Inamura K, Kufe D. Colony-stimulating factor 1-induced Na$^+$ influx into human monocytes involves activation of a pertussis toxin-sensitive GTP-binding protein. J Biol Chem 1988; 263:14093-14098.

92. Nishimoto I, Murayama Y, Katada T et al. Possible direct linkage of insulin-like growth factor-II receptor with guanine nucleotide-binding proteins. J Biol Chem 1989; 264:14029-14038.

93. Luttrell L, Kilgour E, Larner J et al. A pertussis toxin-sensitive G-protein mediates some aspects of insulin action in BC3H-1 murine monocytes. J Biol Chem 1990; 265:16873-16879.

94. Liang M, Garrison JC. The epidermal growth factor is coupled to a pertussis toxin-sensitive guanine nucleotide regulatory protein in rat hepatocytes. J Biol Chem 1991; 266:13342-13349.

95. Okamoto T, Katada T, Murayama Y et al. A simple structure encodes G protein-activating function of the IGFII/mannose 6-phosphate receptor. Cell 1990; 62:709-717.

96. Nishimoto I, Ogata E, Okamoto T. Guanine nucleotide-binding protein interacting but unstimulating sequence located in insulin-like growth factor II receptor. Its autoinhibitory characteristics and structural determinants. J Biol Chem 1991; 266:12747-12751.

97. Kyte J, Doolittle RF. A simple method for displaying the hydropathic character of a protein. J Mol Biol 1982; 157:105-132.

98. Chabre M. Trigger and amplification mechanisms in visual phototransduction. Annu Rev Biophys Biophys Chem 1985; 14:331-360.

99. Dixon RAF, Kobilka BK, Strader DJ et al. Cloning of the gene and cDNA for the mammalian β-adrenergic receptor and homology with rhodopsin. Nature 1986; 321:75-79.

100. Henderson R, Unwin PNT. Three-dimensional model of purple membrane obtained by electron microscopy. Nature 1975;

257:28-32.

101. Henderson R, Baldwin JM, Ceska TA et al. Model for the structure of bacteriorhodopsin based on high-resolution electron cryo-microscopy. J Mol Biol 1990; 213:899-929.

102. Caspar DLD. Bacteriorhodopsin - at last! Nature 1990; 345:666-667.

103. Findlay J, Eliopoulos E. Three-dimensional modelling of G protein-linked receptors. Trends Pharmacol Sci 1990; 11:492-499.

104. Attwood TK, Eliopoulos E, Findlay JBC. Multiple sequence alignment of protein families showing low sequence homology: a methodological approach using database pattern-matching discriminators for G-protein-linked receptors. Gene 1991; 98:153-159.

105. Applebury ML, Hargrave PA. Molecular biology of the visual pigments. Vision Res 1986; 26:1881-1895.

106. Wang H, Lipfert L, Malbon CC et al. Site-directed anti-peptide antibodies define the topography of the β-adrenergic receptor. J Biol Chem 1989; 264:14424-14431.

107. Donnelly D, Johnson MS, Blundell TL et al. An analysis of the periodicity of conserved residues in sequence alignments of G-protein coupled receptors. Implications for the three-dimensional structure. FEBS Lett 1989; 251:109-116.

108. Hibert MF, Trumpp-Kallmeyer S, Bruinvels A et al. Three-dimensional models of neurotransmitter G-binding protein-coupled receptors. Mol Pharmacol 1991; 40:8-15.

109. Dahl SG, Edvardsen Ø, Sylte I. Molecular dynamics of dopamine at the $D_2$ receptor. Proc Natl Acad Sci USA 1991; 88:8111-8115.

110. Pardo L, Ballesteros JA, Osman R et al. On the use of the transmembrane domain of bacteriorhodopsin as a template for modeling the three-dimensional structure of guanine nucleotide-binding regulatory protein-coupled receptors. Proc Natl Acad Sci USA 1992; 89:4009-4012.

111. Trumpp-Kallmeyer S, Hoflack J, Bruinvels A et al. Modeling of G-protein-coupled receptors: application to dopamine, adrenaline, serotonin, acetylcholine, and mammalian opsin receptors. J Med Chem 1992; 35:3448-3462.

112. Cronet P, Sander C, Vriend G. Modeling of transmembrane seven helix bundles. Protein Eng 1993; 6:59-64.

113. Sylte I, Edvardsen Ø, Dahl SG. Molecular dynamics of the 5-HT$_{1a}$ receptor and ligands. Protein Eng 1993; 6:691-700.

114. Brann MR, Klimkowski VJ, Ellis J. Structure/function relationships of muscarinic acetylcholine receptors. Life Sci 1993; 52:405-412.

115. Baldwin JM. The probable arrangement of the helices in G protein-coupled receptors. EMBO J 1993; 12:1693-1703.

116. Schertler GFX, Villa C, Henderson R. Projection structure of rhodopsin. Nature 1993; 362:770-772.

117. Hoflack J, Trumpp-Kallmeyer S, Hibert M. Re-evaluation of bacteriorhodopsin as a model for G protein-coupled receptors. Trends Pharmacol Sci 1994; 15:7-9.

118. Hepler JR, GIlman AG. G proteins. Trends Biochem Sci 1992; 17:383-387.

119. Hille B. G protein-coupled mechanisms and nervous signaling. Neuron 1992; 9:187-195.

120. Masu M, Tanabe Y, Tsuchida K et al. Sequence and expression of a metabotropic glutamate receptor. Nature 1991; 349:760-765.

121. Dias JA. Recent progress in structure-function and molecular analyses of the pituitary/placental glycoprotein hormone receptors. Biochim Biophys Acta 1992; 1135:278-294.

122. Seeman P. Receptor amino acid sequences of G-linked receptors. First Edition. Toronto: University of Toronto, 1992.

123. Abe K, Kusakabe Y, Tanemura K et al. Multiple genes for G protein-coupled receptors and their expression in lingual epithelia. FEBS Lett 1993; 316:253-256.

124. Cone RD, Mountjoy KG. Cloning and functional characterization of the human adrenocorticotropin receptor. Cellular and Molecular Biology of the Adrenal Cortex 1992; 222:27-40.

125. Brown EM, Gamba G, Riccardi D et al. Cloning and characterization of an extracellular $Ca^{2+}$-sensing receptor from bovine parathyroid. Nature 1993; 366:575-580.

126. Tanabe Y, Masu M, Ishii T et al. A family of metabotropic glutamate receptors. Neuron 1992; 8:169-179.

127. Pin J-P, Waeber C, Prezeau L et al. Alternative splicing generates metabotropic glutamate receptors inducing different patterns of calcium release in *Xenopus* oocytes. Proc Natl Acad Sci USA 1992; 89:10331-10335.

128. Abe T, Sugihar H, Nawa H et al. Molecular characterization of a novel metabotropic glutamate receptor mGluR5 coupled to inositol phosphate/$Ca^{2+}$ signal transduction. J Biol Chem 1992; 267:13361-13368.

129. Nakajima Y, Iwakabe H, Akazawa C et al. Molecular characterization of a novel retinal metabotropic glutamate receptor mGluR6 with a high agonist selectivity for L-2-amino-4-phosphonobutyrate. J Biol Chem 1993; 268:11868-11873.

130. Okamoto N, Hori S, Akazawa C et al. Molecular characterization of a new metabotropic glutamate receptor mGluR7 coupled to inhibitory cyclic AMP signal transduction. J Biol Chem 1994; 269:1231-1236.

131. Furlong TJ, Pierce KD, Selbie LA et al. Molecular characterization of a human brain adenosine $A_2$ receptor. Mol Brain Res 1992; 15:62-66.

132. Pierce KD, Furlong TJ, Selbie LA et al. Molecular cloning and expression of an adenosine A2b receptor from human brain. Biochem Biophys Res Commun 1992; 187:86-93.

133. Peralta EG, Ashkenazi A, Winslow JW et al. Distinct primary structures, ligand-binding properties, and tissue-specific expression of four human muscarinic acetylcholine receptors. EMBO J 1987; 6:3923-3929.

134. Boulay F, Tardif M, Brouchon L et al. The human *N*-formylpeptide receptor. Characterization of two cDNA isolates and evidence for a new subfamily of G-protein-coupled receptors. Biochemistry 1990; 29:11123-11133.

135. Gerard NP, Gerard C. The chemotactic receptor for human C5a anaphylatoxin. Nature 1991; 349:614-617.

136. Boulay F, Mary L, Tardif M et al. Expression cloning of a receptor for C5a anaphylatoxin on differentiated HL-60 cells. Biochemistry 1991; 30:2993-2999.

137. Cotecchia S, Schwinn DA, Randall RR et al. Molecular cloning and expression of the cDNA for the hamster $\alpha_1$-adrenergic receptor. Proc Natl Acad Sci USA 1988; 85:7159-7163.

138. Selbie LA, Hayes G, Shine J. The major dopamine D2 receptor: molecular analysis of the human $D2_A$ subtype. DNA Cell Biol 1989; 8:683-689.

139. Brooks J, Taylor PL, Saunders PTK et al. Cloning and sequencing of the sheep pituitary gonadotropin-releasing hormone receptor and changes in expression of its mRNA during the estrous cycle. Mol Cell Endocrinol 1993; 94:R23-R27.

140. Nagayama Y, Rapoport B. The thyrotropin receptor 25 years after its discovery: new insight after its molecular cloning. Mol Endocrinol 1992; 6:145-156.

141. Segaloff DL, Ascoli M. The lutropin/choriogonadotropin receptor...4 years later. Endocr Rev 1993; 14:324-347.

142. Sprengel R, Braun T, Nikolics K et al. The testicular receptor for follicle stimulating hormone: structure and functional expression of cloned cDNA. Mol Endocrinol 1990; 4:525-530.

143. Hubbard SC, Ivatt RJ. Synthesis and processing of asparagine-linked oligosaccharides. Annu Rev Biochem 1981; 50:555-583.

144. Kornfeld R, Kornfeld S. Assembly of asparagine-linked oligosaccharides. Annu Rev Biochem 1985; 54:631-664.

145. Townsend-Nicholson A, Shine J. Molecular cloning and characterisation of a human brain $A_1$ adenosine receptor cDNA. Mol Brain Res 1992; 16:365-370.

146. Zeng D, Harrison JK, D'Angelo DD et al. Molecular characterization of a rat $\alpha_{2B}$-adrenergic receptor. Proc Natl Acad Sci USA 1990; 87:3102-3106.

147. Lomasney JW, Lorenz W, Allen LF et al. Expansion of the $\alpha_2$-adrenergic receptor family: cloning and characterization of a human $\alpha_2$-adrenergic receptor subtype, the gene for which is located on chromosome 2. Proc Natl Acad Sci USA 1990; 87:5094-5098.

148. Kursar JD, Nelson DL, Wainscott DB et al. Molecular cloning, functional expression, and pharmacological characterization of a

novel serotonin receptor (5-hydroxy-tryptamine$_{2F}$) from rat stomach fundus. Mol Pharmacol 1992; 42:549-557.

149. Nakamura M, Honda Z, Izumi T et al. Molecular cloning and expression of platelet-activating factor receptor from human leukocytes. J Biol Chem 1991; 266:20400-20405.

150. Usdin TB, Mezey E, Button DC et al. Gastric inhibitory polypeptide receptor, a member of the secretin-vasoactive intestinal polypeptide receptor family, is widely distributed in peripheral organs and the brain. Endocrinology 1993; 133:2861-2870.

151. Gantz I, Miwa H, Konda Y et al. Molecular cloning, expression, and gene localization of a fourth melanocortin receptor. J Biol Chem 1993; 268:15174-15179.

152. Webb TE, Simon J, Krishek BJ et al. Cloning and functional expression of a brain G-protein-coupled ATP receptor. FEBS Lett 1993; 324:219-225.

153. Ngai J, Dowling MM, Buck L et al. The family of genes encoding odorant receptors in the channel catfish. Cell 1993; 72:657-666.

154. Hargrave PA. The amino-terminal tryptic peptide of bovine rhodopsin. A glycopeptide containing two sites of oligosaccharide attachment. Biochim Biophys Acta 1977; 492:83-94.

155. van Galen PJM, Stiles GL, Michaels G et al. Adenosine A$_1$ and A$_2$ receptors: structure-function relationships. Med Res Rev 1992; 12:423-471.

156. Sawutz DG, Lanier SM, Warren CD et al. Glycosylation of the mammalian α$_1$-adrenergic receptor by complex type N-linked oligosaccharides. Mol Pharmacol 1987; 32:565-571.

157. O'Dowd BF, Lefkowitz RJ, Caron MG. Structure of the adrenergic and related receptors. Annu Rev Neurosci 1989; 12:67-83.

158. van Koppen CJ, Nathanson NM. Site-directed mutagenesis of the m2 muscarinic acetylcholine receptor. analysis of the role of N-glycosylation in receptor expression and function. J Biol Chem 1990; 265:20887-20892.

159. Desarnaud F, Marie J, Lombard C et al. Deglycosylation and fragmentation of purified rat liver angiotensin II receptor: application to the mapping of hormone-binding domains. Biochem J 1993; 289:289-297.

160. Von Heijne G. The signal peptide. J Membr Biol 1990; 115:195-201.

161. Saito Y, Mizuno T, Itakura M et al. Primary structure of bovine endothelin ET$_B$ receptor and identification of signal peptidase and metal proteinase cleavage sites. J Biol Chem 1991; 266:23433-23437.

162. Haendler B, Hechler U, Schleuning W-D. Molecular cloning of human endothelin (ET) receptors ET$_A$ and ET$_B$. J Cardiovasc Pharmacol 1992; 20 (Suppl 12):S1-S4.

163. Karne S, Jayawickreme CK, Lerner MR. Cloning and characterization of an endothelin-3 specific receptor (ET$_C$ receptor) from *Xenopus laevis* dermal melanophores. J Biol Chem 1993; 268:19126-19133.

164. Zhong C, Hayzer DJ, Corson MA et al. Molecular cloning of the rat vascular smooth muscle thrombin receptor. Evidence for *in vitro* regulation by basic fibroblast growth factor. J Biol Chem 1992; 267:16975-16979.

165. Ishihara T, Nakamura S, Kaziro Y et al. Molecular cloning and expression of a cDNA encoding the secretin receptor. EMBO J 1991; 10:1635-1641.

166. Sexton PM, Houssami S, Hilton JM et al. Identification of brain isoforms of the rat calcitonin receptor. Mol Endocrinol 1993; 7:815-821.

167. Chen R, Lewis KA, Perrin MH et al. Expression cloning of a human corticotropin-releasing-factor receptor. Proc Natl Acad Sci USA 1993; 90:8967-8971.

168. Perrin MH, Donaldson CJ, Chen R et al. Cloning and functional expression of a rat brain corticotropin releasing factor (CRF) receptor. Endocrinology 1993; 133:3058-3061.

169. Chang C-P, Pearse RV, O'Connell S et al. Identification of a seven transmembrane helix receptor for corticotropin-releasing factor and sauvagine in mammalian brain. Neuron 1993; 11:1187-1195.

170. Svoboda M, Ciccarelli E, Tastenoy M et al. Small introns in a hepatic cDNA encoding a new glucagon-like peptide 1-type recep-

tor. Biochem Biophys Res Commun 1993; 191:479-486.

171. Dillon JS, Tanizawa Y, Wheeler MB et al. Cloning and functional expression of the human glucagon-like peptide-1 (GLP-1) receptor. Endocrinology 1993; 133:1907-1910.

172. Mayo KE. Molecular cloning and expression of a pituitary-specific receptor for growth hormone-releasing hormone. Mol Endocrinol 1992; 6:1734-1744.

173. Lin C, Lin S-C, Chang C-P et al. Pit-1-dependent expression of the receptor for growth hormone releasing factor mediates pituitary cell growth. Nature 1992; 360:765-768.

174. Hsiung HM, Smith DP, Zhang X-Y et al. Structure and functional expression of a complementary DNA for porcine growth hormone-releasing hormone receptor. Neuropeptides 1993; 25:1-10.

175. Schipani E, Karga H, Karpalis AC et al. Identical complementary deoxyribonucleic acids encode a human renal and bone parathyroid hormone (PTH)/PTH-related peptide receptor. Endocrinology 1993; 132:2157-2165.

176. Spengler D, Waeber C, Pantaloni C et al. Differential signal transduction of five splice variants of the PACAP receptor. Nature 1993; 365:170-175.

177. Sreedharan SP, Patel DR, Huang J-X et al. Cloning and functional expression of a human neuroendocrine vasoactive intestinal peptide receptor. Biochem Biophys Res Commun 1993; 193:546-553.

178. Battey JF, Way JM, Corjay MH et al. Molecular cloning of the bombesin/gastrin-releasing peptide receptor from Swiss 3T3 cells. Proc Natl Acad Sci USA 1991; 88:395-399.

179. Findlay JBC, Pappin DJC. The opsin family of proteins. Biochem J 1986; 238:625-642.

180. Oprian DD. The ligand-binding domain of rhodopsin and other G protein-linked receptors. J Bioenerg Biomembr 1992; 24:211-217.

181. Takahashi K, Tsuchida K, Tanabe Y et al. Role of the large extracellular domain of metabotropic glutamate receptors in ago-

nist selectivity determination. J Biol Chem 1993; 268:19341-19345.

182. Iismaa TP, Shine J. G protein-coupled receptors. Curr Opin Cell Biol 1992; 4:195-202.

183. Nakanishi S, Nakajima Y, Yokota Y. Signal transduction and ligand-binding domains of the tachykinin receptors. Regul Pept 1993; 46:37-42.

184. Lefkowitz RJ, Caron MG. Adrenergic receptors: models for the study of receptors coupled to guanine nucleotide regulatory proteins. J Biol Chem 1988; 263:4993-4996.

185. Strader CD, Sigal IS, Dixon RAF. Mapping the functional domains of the β-adrenergic receptor. Am J Respir Cell Mol Biol 1989; 1:81-86.

186. Strader CD, Sigal IS, Dixon RAF. Structural basis of β-adrenergic function. FASEB J 1989; 3:1825-1832.

187. Dohlman HG, Thorner MG, Caron MG et al. Model systems for the study of seven-transmembrane-segment receptors. Annu Rev Biochem 1991; 60:653-688.

188. Kobilka B. Adrenergic receptors as models for G protein-coupled receptors. Annu Rev Neurosci 1992; 15:87-114.

189. Strosberg AD. Structure/function relationships of proteins belonging to the family of receptors coupled to GTP-binding proteins. Eur J Biochem 1991; 196:1-10.

190. Strosberg AD. Structure, function, and regulation of adrenergic receptors. Prot Sci 1993; 2:1198-1209.

191. Huganir RL, Greengard P. Regulation of neurotransmitter desensitization by protein phosphorylation. Neuron 1990; 5:555-567.

192. Dixon RAF, Sigal IS, Candelore MR et al. Structural features required for ligand binding to the β-adrenergic receptor. EMBO J 1987; 6:3269-3275.

193. Karnik SS, Sakmar TP, Chen HB et al. Cysteine residues 110 and 187 are essential for the formation of correct structure in bovine rhodopsin. Proc Natl Acad Sci USA 1988; 85:8459-8463.

194. Fraser CM. Site-directed mutagenesis of β-adrenergic receptors: identification of conserved cysteine residues that independently affect ligand binding and receptor activa-

tion. J Biol Chem 1989; 264:9266-9270.

195. Hulme EC, Birdsall NJ, Buckley NJ. Muscarinic receptor subtypes. Annu Rev Pharmacol Toxicol 1990; 30:633-673.

196. Noda K, Saad Y, Graham RM et al. The high affinity state of the $\beta_2$-adrenergic receptor requires unique interaction between conserved and nonconserved extracellular loop cysteines. J Biol Chem 1994; 269:6743-6752.

197. Ovchinnikov YA, Abdulalv NG, Bogachuk AS. Two adjacent cysteine residues in the C-terminal cytoplasmic fragment of bovine rhodopsin are palmitoylated. FEBS Lett 1988; 230:1-5.

198. O'Dowd BF, Hnatowich M, Caron MG et al. Palmitoylation of the human $\beta$2-adrenergic receptor. Mutation of Cys$^{341}$ in the carboxyl tail leads to an uncoupled nonpalmitoylated form of the receptor. J Biol Chem 1989; 264:7564-7569.

199. Probst WC, Snyder LA, Schuster DI et al. Sequence alignment of the G-protein coupled receptor superfamily. DNA Cell Biol 1992; 11:1-20.

200. Kobilka BK, Dixon RAF, Frielle T et al. cDNA for the human $\beta_2$-adrenergic receptor: a protein with multiple membrane-spanning domains and encoded by a gene whose chromosomal location is shared with that of the receptor for platelet-derived growth factor. Proc Natl Acad Sci USA 1987; 84:46-50.

201. Chung F-Z, Lentes KU, Gocayne J et al. Cloning and sequence analysis of the human $\beta$-adrenergic receptor. Evolutionary relationship to rodent and avian $\beta$-receptors and porcine muscarinic receptors. FEBS Lett 1987; 211:200-206.

202. Strader CD, Sigal IS, Register RB et al. Identification of residues required for ligand binding to the $\beta$-adrenergic receptor. Proc Natl Acad Sci USA 1987; 84:4384-4388.

203. Wess J, Nanavati S, Vogel Z et al. Functional role of proline and tryptophan residues highly conserved among G protein-coupled receptors studied by mutational analysis of the m3 muscarinic receptor. EMBO J 1993; 12:331-338.

204. Burbach JPH, Meijer OC. The structure of neuropeptide receptors. Eur J Pharmacol 1992; 227:1-18.

205. Wang Z, Wang H, Ascoli M. Mutation of a highly conserved acidic residue present in the second intracellular loop of G-protein-coupled receptors does not impair hormone binding or signal transduction of the luteinizing hormone/chorionic gonadotropin receptor. Mol Endocrinol 1993; 7:85-93.

206. Kimura T, Tanizawa O, Mori K et al. Structure and expression of a human oxytocin receptor. Nature 1992; 356:526-529. Erratum Nature 357:176.

207. Masu YK, Nakayama K, Tamaki Y et al. cDNA cloning of bovine substance-K receptor through oocyte expression system. Nature 1987; 329:836-838.

208. Honda Z, Nakamura M, Miki I et al. Cloning by functional expression of platelet-activating factor receptor from guinea-pig lung. Nature 1991; 349:342-346.

209. Gluzman Y. SV-40 transformed simian cells support the replication of early SV40 mutants. Cell 1981; 23:175-182.

210. Seed B. An LFA-3 cDNA encodes a phospholipid-linked membrane protein homologous to its receptor CD2. Nature 1987; 329:840-842.

211. Seed B, Aruffo A. Molecular cloning of the CD2 antigen, the T-cell erythrocyte receptor, by a rapid immunoselection procedure. Proc Natl Acad Sci USA 1987; 84:3365-3369.

212. Xie G-X, Miyajima A, Goldstein A. Expression cloning of cDNA encoding a seven-helix receptor from human placenta with affinity for opioid ligands. Proc Natl Acad Sci USA 1992; 89:4124-4128. Erratum Proc Natl Acad Sci USA 89:7287.

213. Harada N, Castle BE, Gorman DM et al. Expression cloning of a cDNA encoding the murine interleukin 4 receptor based on ligand binding. Proc Natl Acad Sci USA 1990; 87:857-861.

214. Holmes WE, Lee J, Kuang W-J et al. Structure and functional expression of a human interleukin-8 receptor. Science 1991; 253:1278-1280.

215. Kieffer BL, Befort K, Gaveriaux-Ruff C et al. The $\delta$-opioid receptor: isolation of a cDNA clone by expression cloning and pharmacological characterization. Proc Natl

Acad Sci USA 1992; 89:12048-12052.

216. Evans CJ, Keith DE, Morrison H et al. Cloning of a delta opioid receptor by functional expression. Science 1992; 258:1952-1955.

217. Barberis C, Seibold A, Ishido M et al. Expression cloning of the human $V_2$ vasopressin receptor. Regul Pept 1993; 45:61-66.

218. McClintock TS, Graminski GF, Potenza MN et al. Functional expression of recombinant G-protein coupled receptors monitored by video imaging of pigment movement in melanophores. Anal Biochem 1993; 209:298-305.

219. Herzog H, Hort YJ, Ball HJ et al. Cloned human neuropeptide Y receptor couples to two different second messenger systems. Proc Natl Acad Sci USA 1992; 89:5794-5798.

220. Perret JJ, Raspe E, Vassart G et al. Cloning and functional expression of the canine anaphylatoxin C5a receptor. Biochem J 1992; 288:911-917.

221. Sunahara RK, Guan H-C, O'Dowd BF et al. Cloning of the gene for a human dopamine $D_5$ receptor with higher affinity for dopamine than $D_1$. Nature 1991; 350:614-619.

222. Ishihara T, Shigemoto R, Mori K et al. Functional expression and tissue distribution of a novel receptor for vasoactive intestinal peptide. Neuron 1992; 8:811-819.

223. Bunzow JR, Van Tol HHM, Grandy DK et al. Cloning and expression of a rat $D_2$ dopamine receptor cDNA. Nature 1988; 336:783-787.

224. Matsuda LA, Lolait SJ, Brownstein MJ et al. Structure of a cannabinoid receptor and functional expression of the cloned cDNA. Nature 1990; 346:561-564.

225. Devane WA, Hanus L, Breuer A et al. Isolation and structure of a brain constituent that binds to the cannabinoid receptor. Science 1992; 258:1946-1949.

226. Felder CC, Briley EM, Axelrod J et al. Anandamide, an endogenous cannabimimetic eicosanoid, binds to the cloned human cannabinoid receptor and stimulates receptor-mediated signal transduction. Proc Natl Acad Sci USA 1993; 90:7656-7660.

227. Libert F, Parmentier M, Lefort A et al. Selective amplification and cloning of four new members of the G protein-coupled receptor family. Science 1989; 244:569-572.

228. Maenhaut C, Van Sande J, Libert F et al. RDC8 codes for an adenosine A2 receptor with physiological constitutive activity. Biochem Biophys Res Commun 1990; 173:1169-1178.

229. Libert F, Schiffmann S, Lefort A et al. The orphan receptor cDNA RDC7 encodes an A1 adenosine receptor. EMBO J 1991; 10:1677-1682.

230. Watson S, Girdlestone D. Receptor & Ion Channel Nomenclature Supplement. Trends Pharmacol Sci 1994; 15 Suppl:1-51.

231. Buck L, Axel R. A novel multigene family may encode odorant receptors: a molecular basis for odor recognition. Cell 1991; 65:175-187.

232. Selbie LA, Townsend-Nicholson A, Iismaa TP et al. Novel G protein-coupled receptors: a gene family of putative human olfactory receptor sequences. Mol Brain Res 1992; 13:159-163.

233. Bao L, Gerard NP, Eddy RL et al. Mapping of genes for the human-C5a receptor (C5AR), human FMLP receptor (FPR), and 2 FMLP receptor homologue orphan receptors (FPRH1, FPRH2) to chromosome 19. Genomics 1992; 13:437-440.

234. Murphy PM, Özçelik T, Kenney RT et al. A structural homologue of the N-formyl peptide receptor. Characterization and chromosome mapping of a peptide chemoattractant receptor family. J Biol Chem 1992; 267:7637-7643.

235. Ahuja SK, Özçelik T, Milatovitch A et al. Molecular evolution of the human interleukin-8 receptor gene cluster. Nature Genetics 1992; 2:31-36.

236. Herzog H, Hort YJ, Shine J et al. Molecular cloning, characterization, and localization of the human homolog to the reported bovine NPY Y3 receptor: lack of NPY binding and activation. DNA Cell Biol 1993; 12:465-471.

237. Harrison JK, Barber CM, Lynch KR. Molecular cloning of a novel rat G-protein-coupled receptor gene expressed prominently in lung, adrenal, and liver. FEBS Lett 1993; 318:17-22.

238. Eva C, Sprengel R. A novel putative G protein-coupled receptor highly expressed in lung and testis. DNA Cell Biol 1993; 12:393-399.

239. Perez DM, Piascik MT, Graham RM. Solution-phase library screening for the identification of rare clones: isolation of an $\alpha_{1D}$-adrenergic receptor cDNA. Mol Pharmacol 1991; 40:876-883.

240. Yasuda K, Raynor K, Kong H et al. Cloning and functional comparison of κ and δ opioid receptors from mouse brain. Proc Natl Acad Sci USA 1993; 90:6736-6740.

241. Fishburn CS, Belleli D, David C et al. A novel short isoform of the $D_3$ dopamine receptor generated by alternative splicing in the third cytoplasmic loop. J Biol Chem 1993; 268:5872-5878.

242. Itokawa M, Arinami T, Futamura N et al. A structural polymorphism of human dopamine D2 receptor. D2($Ser^{311}$-Cys). Biochem Biophys Res Commun 1993; 196:1369-1375.

243. Gingrich JA, Caron MG. Recent advances in the molecular biology of dopamine receptors. Annu Rev Neurosci 1993; 16:299-321.

244. O'Dowd BF. Structures of dopamine receptors. J Neurochem 1993; 60:804-816.

245. Giros B, Sokoloff P, Martres MP et al. Alternative splicing directs the expression of two $D_2$ dopamine receptor isoforms. Nature 1989; 342:923-926.

246. Monsma FJ, McVittie LD, Gerfen CR et al. Multiple $D_2$ dopamine receptors produced by alternative RNA splicing. Nature 1989; 342:926-928.

247. Dal Toso R, Sommer B, Ewert M et al. The dopamine $D_2$ receptor: two molecular forms generated by alternative splicing. EMBO J 1989; 8:4025-4034.

248. Montmayeur J-P, Borrelli E. Transcription mediated by a cAMP-responsive promoter element is reduced upon activation of dopamine $D_2$ receptors. Proc Natl Acad Sci USA 1991; 88:3135-3139.

249. Hayes G, Biden TJ, Selbie LA et al. Structural subtypes of the dopamine D2 receptor are functionally distinct: expression of the cloned D2A and D2B subtypes in a heterologous cell line. Mol Endocrinol 1992; 6:920-926.

250. Namba T, Sugimoto Y, Negishi M et al. Alternative splicing of C-terminal tail of prostaglandin E receptor subtype EP3 determines G-protein specificity. Nature 1993; 365:166-170.

251. Irie A, Sugimoto Y, Namba T et al. Third isoform of the prostaglandin-E-receptor $EP_3$ subtype with different C-terminal tail coupling to both stimulation and inhibition of adenylate cyclase. Eur J Biochem 1993; 217:313-318.

252. Levy NS, Bakalyar HA, Reed RR. Signal transduction in olfactory neurons. J Steroid Biochem 1991; 39:633-637.

253. Parmentier M, Libert F, Schurmans S et al. Expression of members of the putative olfactory receptor gene family in mammalian germ cells. Nature 1992; 355:453-455.

254. Matsuoka I, Mori T, Aoki J et al. Identification of novel members of G-protein coupled receptor superfamily expressed in bovine taste tissue. Biochem Biophys Res Commun 1993; 194:504-511.

255. Salesse R, Remy JJ, Levin JM et al. Towards understanding the glycoprotein hormone receptors. Biochimie 1991; 73:109-120.

256. Lin HY, Harris TL, Flannery MS et al. Expression cloning of an adenylate cyclase-coupled calcitonin receptor. Science 1991; 254:1022-1024.

257. Jelinek LJ, Lok S, Rosenberg GB et al. Expression cloning and signaling properties of the rat glucagon receptor. Science 1993; 259:1614-1616.

258. Abou-Samra A-B, Jüppner H, Force T et al. Expression cloning of a common receptor for parathyroid hormone and parathyroid hormone-related peptide from rat osteoblast-like cells: a single receptor stimulates intracellular accumulation of both cAMP and inositol triphosphates and increases intracellular free calcium. Proc Natl Acad Sci USA 1992; 89:2732-2736.

259. Segre GV, Goldring SR. Receptors for secretin, calcitonin, parathyroid hormone (PTH)/PTH-related peptide, vasoactive intestinal peptide, glucagonlike peptide 1, growth hormone-releasing hormone, and glucagon belong to a newly discovered G-

protein-linked receptor family. Trends Endocrinol Metab 1993; 4:309-314.

260. Reagan JD. Expression cloning of an insect diuretic hormone receptor. A member of the calcitonin/secretin receptor family. J Biol Chem 1994; 269:9-12.

261. Taylor EW, Agarwal A. Sequence homology between bacteriorhodopsin and G-protein coupled receptors: exon shuffling or evolution by duplication? FEBS Lett 1993; 325:161-166.

262. Khorana HG. Rhodopsin, photoreceptor of the rod cell. J Biol Chem 1992; 267:1-4.

263. Yokoyama S, Isenberg KE, Wright AF. Adaptive evolution of G-protein coupled receptor genes. Mol Biol Evol 1989; 6:342-353.

264. Klein PS, Sun TJ, Saxe CL et al. A chemoattractant receptor controls development in *Dictyostelium discoideum*. Science 1988; 241:1467-1472.

265. Burkholder AC, Hartwell LH. The yeast α-factor receptor: structural properties deduced from the sequence of the *STE2* gene. Nucleic Acids Res 1985; 13:8463-8475.

266. Marsh L, Herskowitz I. STE2 protein of *Saccharomyces kluyveri* is a member of the rhodopsin/β-adrenergic receptor family and is responsible for recognition of the peptide ligand α factor. Proc Natl Acad Sci USA 1988; 85:3855-3859.

267. Nakayama N, Miyajima A, Arai K. Nucleotide sequence of *STE2* and *STE3*, cell type-specific sterile genes from *Saccharomyces cerevisiae*. EMBO J 1985; 4:2643-2648.

268. Ohno S. Evolution by gene duplication. New York: Springer, 1970.

269. Brosius J. Retroposons - seeds of evolution. Science 1991; 251:753.

270. Schaffner W, Kunz G, Daetwyler H et al. Genes and spacers of cloned sea urchin histone DNA analyzed by sequencing. Cell 1978; 14:655-671.

271. Nagata S, Mantei N, Weissmann C. The structure of one of the eight or more distinct chromosomal genes for human interferon-alpha. Nature 1980; 287:401-408.

272. Lawn RM, Adelman J, Franke AE et al. Human fibroblast interferon gene lacks introns. Nucleic Acids Res 1981; 9:1045-1052.

273. Ninomiya Y, Gordon M, van der Rest M et al. The developmentally regulated type X collagen gene contains a long open reading frame without introns. J Biol Chem 1986; 261:5041-5050.

274. Song I, Brown DR, Wiltshire RN et al. The human gastrin/cholecystokinin type B receptor gene: alternative splice donor site in exon 4 generates two variant mRNAs. Proc Natl Acad Sci USA 1993; 90:9085-9089.

275. Hosoda K, Nakao K, Tamura N et al. Organization, structure, chromosomal assignment, and expression of the gene encoding the human endothelin-A receptor. J Biol Chem 1992; 267:18797-18804.

276. Arai H, Nakao K, Takaya K et al. The human endothelin-B receptor gene. Structural organization and chromosomal assignment. J Biol Chem 1993; 268:3463-3470.

277. Eva C, Oberto A, Sprengel R et al. The murine NPY-1 receptor gene. Structure and delineation of tissue-specific expression. FEBS Lett 1992; 314:285-288.

278. Herzog H, Baumgartner M, Vivero C et al. Genomic organization, localization, and allelic differences in the gene for the human neuropeptide Y Y1 receptor. J Biol Chem 1993; 268:6703-6707.

279. Heckert LL, Daley IJ, Griswold MD. Structural organization of the follicle-stimulating hormone receptor gene. Mol Endocrinol 1992; 6:70-80.

280. Ramarao CS, Kincade Denker JM, Perez D et al. Genomic organization and expression of the human $\alpha_{1B}$-adrenergic receptor. J Biol Chem 1992; 267:21936-21945.

281. Granneman JG, Lahners KN, Rao DD. Rodent and human $\beta_3$-adrenergic receptor genes contain an intron within the protein-coding block. Mol Pharmacol 1992; 42:964-970.

282. Chen K, Yang W, Grimsby J et al. The human 5-HT$_2$ receptor is encoded by a multiple intron-exon gene. Mol Brain Res 1992; 14:20-26.

283. Stam NJ, Van Huizen F, Van Alebeek C et al. Genomic organization, coding sequence and functional expression of human 5-HT$_2$ and 5-HT$_{1A}$ receptor genes. Eur J Pharmacol 1992; 227:153-162.

284. Shen Y, Monsma FJ, Metcalf MA et al. Molecular cloning and expression of a 5-hydroxytryptamine$_7$ serotonin receptor subtype. J Biol Chem 1993; 24:18200-18204.

285. Ruat M, Traiffort E, Leurs R et al. Molecular cloning, characterization, and localization of a high-affinity serotonin receptor (5-HT$_7$) activating cAMP formation. Proc Natl Acad Sci USA 1993; 90:8547-8551.

286. Julius D. Molecular biology of serotonin receptors. Annu Rev Neurosci 1991; 14:335-360.

287. Kobilka BK, Frielle T, Dohlman HG et al. Delineation of the intronless nature of the genes for the human and hamster β$_2$-adrenergic receptor and their putative promoter regions. J Biol Chem 1987; 262:7321-7327.

288. Buck LB. The olfactory multigene family. Curr Opin Neurobiol 1992; 2:467-473.

289. Curnow KM, Pascoe L, White PC. Genetic analysis of the human type-1 angiotensin II receptor. Mol Endocrinol 1992; 6:1113-1118.

290. Takeuchi K, Alexander RW, Nakamura Y et al. Molecular structure and transcriptional function of the rat vascular AT$_{1a}$ angiotensin receptor gene. Circulation Res 1993; 73:612-621.

291. Ichiki T, Herold CL, Kambayashi Y et al. Cloning of the cDNA and the genomic DNA of the mouse angiotensin II type 2 receptor. Biochim Biophys Acta 1994; 1189:247-250.

292. Eggerickx D, Raspe E, Bertrand D et al. Molecular cloning, functional expression and pharmacological characterization of a human bradykinin B2 receptor gene. Biochem Biophys Res Commun 1992; 187:1306-1313.

293. Peralta EG, Winslow JW, Peterson GL et al. Primary structure and biochemical properties of an M$_2$ muscarinic receptor. Science 1987; 236:600-605.

294. Murphy PM, Tiffany HL, McDermott D et al. Sequence and organization of the human N-formyl peptide receptor-encoding gene. Gene 1993; 133:285-290.

295. Mutoh H, Bito H, Minami M et al. Two different promoters direct expression of two distinct forms of mRNAs of human platelet-activating factor receptor. FEBS Lett 1993; 322:129-134.

296. Ren H, Stiles GL. Characterization of the human A$_1$ adenosine receptor gene. Evidence for alternative spicing. J Biol Chem 1994; 269:3104-3110.

297. Cheng H-F, Su Y-M, Chang K-J. Alternative transcript of the nonselective-type endothelin receptor from rat brain. Mol Pharmacol 1993; 44:533-538.

298. Nüsing RM, Hirata M, Kakizuka A et al. Characterization and chromosomal mapping of the human thromboxane A$_2$ receptor gene. J Biol Chem 1993; 268:25253-25259.

299. Montmayeur JP, Bausero P, Amlaiky N et al. Differential expression of the mouse D$_2$ dopamine receptor isoforms. FEBS Lett 1991; 278:239-243.

300. Vanetti M, Vogt G, Höllt V. The two isoforms of the mouse somatostatin receptor (mSSTR2A and mSSTR2B) differ in coupling efficiency to adenylate cyclase and in agonist-induced receptor desensitization. FEBS Lett 1993; 331:260-266.

301. Sugimoto U, Negishi M, Hayashi Y et al. Two isoforms of the EP$_3$ receptor with different carboxyl-terminal domains. Identical ligand binding properties and different coupling with G$_i$ proteins. J Biol Chem 1993; 268:2712-2718.

302. Fong TM, Anderson SA, Yu H et al. Differential activation of intracellular effector by two isoforms of human neurokinin-1 receptor. Mol Pharmacol 1992; 41:24-30.

303. de la Peña P, Delgado LM, del Camino D et al. Two isoforms of the thyrotropin-releasing hormone receptor generated by alternative splicing have indistinguishable functional properties. J Biol Chem 1992; 267:25703-25708.

304. Charo IF, Myers SJ, Herman A et al. Molecular cloning and functional expression of two monocyte chemoattractant protein 1 receptors reveals alternative splicing of the carboxyl-terminal tails. Proc Natl Acad Sci USA 1994; 91:2752-2756.

305. Breyer RM, Emeson RB, Tarng J-L et al. Alternative splicing generates multiple isoforms of a rabbit prostaglandin E$_2$ receptor. J Biol Chem 1994; 269:6163-6169.

306. Pagliusi S, Chollet-Daemerius A, Losberger C et al. Characterization of a novel exon

within the D3 receptor gene giving rise to
an mRNA isoform expressed in rat brain.
Biochem Biophys Res Commun 1993;
194:465-471.
307. Giros B, Martres M-P, Pilon C et al. Shorter
variants of the D$_3$ dopamine receptor
produced through various patterns of alternative splicing. Biochem Biophys Res
Commun 1991; 176:1584-1592.
308. Snyder LA, Roberts JL, Sealfon SC. Alternative transcripts of the rat and human
dopamine D3 receptor. Biochem Biophys
Res Commun 1991; 180:1031-1035.
309. Nagai Y, Ueno S, Saeki Y et al. Expression
of the D3 dopamine receptor gene and a
novel variant transcript generated by alternative splicing in human peripheral
blood lymphocytes. Biochem Biophys Res
Commun 1993; 194:368-374.
310. Khan H, Yarney TA, Sairam MR. Cloning
of alternately spliced mRNA transcripts
coding for variants of ovine testicular
follitropin receptor lacking the G protein
coupling domains. Biochem Biophys Res
Commun 1993; 190:888-894.
311. Loosfelt H, Misrahi M, Atger M et al. Cloning and sequencing of porcine LH-hCG
receptor cDNA: variants lacking transmembrane domain. Science 1989; 245:525-528.
312. Tsai-Morris CH, Buczko E, Wang W et al.
Intronic nature of the rat luteinizing hormone receptor gene defines a soluble receptor subspecies with hormone binding activity. J Biol Chem 1990; 265:19385-19388.
313. Tilly JL, Aihara T, Nishimori K et al.
Expression of recombinant human follicle-
stimulating hormone receptor: species-spe-
cific ligand binding, signal transduction, and
identification of multiple ovarian messenger ribonucleic acid transcripts. Endocrinology 1992; 131:799-806.
314. O'Shaughnessy PJ, Dudley K. Discrete splicing alternatives in mRNA encoding the
extracellular domain of the testis FSH receptor in the normal and hypogonadal (hpg)
mouse. J Mol Endocrinol 1993; 10:363-366.
315. Minegish T, Nakamura K, Takakura Y et
al. Cloning and sequencing of human LH/
hCG receptor cDNA. Biochem Biophys Res
Commun 1990; 172:1049-1054.
316. Catalano M, Nobile M, Novelli E et al.

Distribution of a novel mutation in the first
exon of the human dopamine D$_4$ receptor
gene in psychotic patients. Biol Psychiatry
1993; 34:459-464.
317. Van Tol HHM, Wu CM, Guan H-C et al.
Multiple dopamine D4 receptor variants in
the human population. Nature 1992;
358:149-152.
318. Lichter JB, Barr CL, Kennedy JL et al. A
hypervariable segment in the human dopamine receptor D4 (DRD4) gene. Hum Mol
Genet 1993; 2:767-773.
319. Civelli O, Bunzow JR, Grandy DK. Molecular diversity of the dopamine receptors.
Annu Rev Pharmacol Toxicol 1993;
32:281-307.
320. Green SA, Cole G, Jacinto M et al. A polymorphism of the human β$_2$-adrenergic receptor within the fourth transmembrane
domain alters ligand binding and functional
properties of the receptor. J Biol Chem
1993; 268:23116-23121.
321. Yang-Feng TL, Xue F, Zhong W et al.
Chromosomal organization of adrenergic
receptor genes. Proc Natl Acad Sci USA
1990; 87:1516-1520.
322. Yang-Feng TL, Han H, Lomasney JW et
al. Localization of the cDNA for an α$_1$-adr-
energic receptor subtype (ADRA1D) to
chromosome band 20p13. Cytogenet Cell
Genet 1994; 66:170-171.
323. Oakey RJ, Caron MG, Lefkowitz RJ et al.
Genomic organization of adrenergic
and serotonin receptors in the mouse: linkage mapping of sequence-related genes provides a method for examining mammalian
chromosome evolution. Genomics 1991;
10:338-344.
324. Copeland NG, Jenkins NA, Gilbert DJ et
al. A genetic linkage map of the mouse:
current applications and future prospects.
Science 1993; 262:57-66.
325. Lolait SJ, O'Carroll A-M, McBride OW et
al. Cloning and characterization of a vasopressin V2 receptor and possible link to
nephrogenic diabetes insipidus. Nature
1992; 357:336-339.
326. Bonner TI. The molecular basis of muscarinic receptor diversity. Trends Neurosci
1989; 12:148-152.
327. Hayes G, Shine J. Dopamine receptor di-

versity. Today's Life Science 1992; 4:16-22.

328. Collis MG, Hourani SMO. Adenosine receptor subtypes. Trends Pharmacol Sci 1993; 14:360-366.

329. Hausdorff WP, Caron MG, Lefkowitz RJ. Turning off the signal: desensitization of β-adrenergic receptor function. FASEB J 1990; 4:2881-2889. Erratum FASEB J 4:3049.

330. Summers RJ, McMartin LR. Adrenoceptors and their second messengers. J Neurochem 1993; 60:10-23.

331. Nakanishi S. Molecular diversity of glutamate receptors and implications for brain function. Science 1992; 258:597-603.

332. Mills A, Duggan MJ. Orphan seven transmembrane domain receptors: reversing pharmacology. Trends Pharmacol Sci 1993; 14:394-396.

333. Rimland J, Xin W, Sweetnam P et al. Sequence and expression of a neuropeptide Y receptor cDNA. Mol Pharmacol 1991; 40:869-875.

334. Jazin EE, Yoo H, Blomqvist AG et al. A proposed bovine neuropeptide Y (NPY) receptor cDNA clone, or its human homologue, confers neither NPY binding sites nor NPY responsiveness on transfected cells. Regul Pept 1993; 47:247-258.

335. Oppenheim JJ, Zachariae COC, Mukaida N et al. Properties of the novel proinflammatory supergene "intercrine" cytokine family. Annu Rev Immunol 1991; 9:617-648.

336. Kelvin DJ, Michiel DF, Johnson JA et al. Chemokines and serpentines: the molecular biology of chemokine receptors. J Leukoc Biol 1993; 54:604-612.

337. Reppert SM, Weaver DR, Stehle JH et al. Molecular cloning of a G protein-coupled receptor that is highly expressed in lymphocytes and proliferative areas of developing brain. Mol Cell Neurosci 1992; 3:206-214.

338. Federsppiel B, Melhado IG, Duncan AMV et al. Molecular cloning of the cDNA and chromosomal localization of the gene for a putative seven-transmembrane segment (7-TMS) receptor isolated from human spleen. Genomics 1993; 16:707-712.

339. Loetscher M, Geiser T, O'Reilly T et al. Cloning of a human seven-transmembrane domain receptor, LESTR, that is highly expressed in leukocytes. J Biol Chem 1994; 269:232-237.

340. Jackson TR, Blair LAC, Marshall J et al. The *mas* oncogene encodes an angiotensin receptor. Nature 1988; 335:437-440.

341. Ross PC, Figler RA, Corjay MH et al. RTA, a candidate G protein-coupled receptor: cloning, sequencing, and tissue distribution. Proc Natl Acad Sci USA 1990; 87:3052-3056.

342. Monnot C, Weber V, Stinnakre J et al. Cloning and functional characterization of a novel *mas*-related gene, modulating intracellular angiotensin II actions. Mol Endocrinol 1991; 5:1477-1487.

343. Cwirla SE, Peters EA, Barrett RW et al. Peptides on phage: a vast library of peptides for identifying ligands. Proc Natl Acad Sci USA 1990; 87:6378-6382.

344. Devlin JJ, Panganiban LC, Devlin PE. Random peptide libraries: a source of specific protein binding molecules. Science 1990; 249:404-406.

345. Scott JK, Smith GP. Libraries of peptides and proteins displayed on filamentous phage. Science 1990; 249:386-390.

346. Smith GP, Scott JK. Libraries of peptides and proteins displayed on filamentous phage. Meth Enzymol 1993; 217:228-257.

347. Mountjoy KG, Robbins LS, Mortrud MT et al. The cloning of a family of genes that encode the melanocortin receptors. Science 1992; 257:1248-1251.

348. Hahn AWA, Jonas U, Buehler FR et al. Identification of a fourth angiotensin $AT_1$ receptor subtype in rat. Biochem Biophys Res Commun 1993; 192:1260-1265.

349. Nakajima M, Mukoyama M, Pratt RE et al. Cloning of cDNA and analysis of the gene for mouse angiotensin II type 2 receptor. Biochem Biophys Res Commun 1993; 197:393-399.

350. Bergsman DJ, Ellis C, Nuthulaganti PR et al. Isolation and expression of a novel angiotensin II receptor from *Xenopus laevis* heart. Mol Pharmacol 1993; 44:277-284.

351. Ji H, Sandberg K, Zhang Y et al. Molecular cloning, sequencing and functional expression of an amphibian angiotensin II receptor. Biochem Biophys Res Commun 1993; 194:756-762.

352. Young D, Waitches G, Birchmeier C et al. Isolation and characterization of a new cellular oncogene encoding a protein with multiple potential transmembrane domains. Cell 1986; 45:711-719.

353. Young D, O'Neill K, Jessell T et al. Characterisation of the rat *mas* oncogene and its high-level expression in the hippocampus and cerebral cortex of rat brain. Proc Natl Acad Sci USA 1988; 85:5339-5342.

354. Ambroz C, Clark AJ, Catt KJ. The *mas* oncogene enhances angiotensin-induced $[Ca^{2+}]_i$ responses in cells with pre-existing angiotensin II receptors. Biochim Biophys Acta 1991; 1133:107-111.

355. Gorn AH, Lin HY, Yamin M et al. Cloning, characterization, and expression of a human calcitonin receptor from an ovarian carcinoma cell line. J Clin Invest 1992; 90:1726-1735.

356. De Weerth A, Pisegna JR, Wank SA. Guinea pig gallbladder and pancreas possess identical CCK-A receptor subtypes: receptor cloning and expression. Am J Physiol 1993; 265:G1116-G1121.

357. Elshourbagy NA, Lee JA, Korman DR et al. Molecular cloning and characterization of the major endothelin receptor subtype in porcine cerebellum. Mol Pharmacol 1991; 41:465-473.

358. Gromoll J, Dankbar B, Sharma RS et al. Molecular cloning of the testicular follicle stimulating hormone receptor of the non human primate *Macaca fascicularis* and identification of multiple transcripts in the testis. Biochem Biophys Res Commun 1993; 196:1066-1072.

359. Yarney TA, Sairam MR, Khan H et al. Molecular cloning and expression of the ovine testicular follicle stimulating hormone receptor. Mol Cell Endocrinol 1993; 93:219-226.

360. MacNeil DJ, Occi JL, Hey PJ et al. Cloning and expression of a human glucagon receptor. Biochem Biophys Res Commun 1994; 198:328-334.

361. Thorens B, Porret A, Bühler L et al. Cloning and functional expression of the human islet GLP-1 receptor. Demonstration that exendin-4 is an agonist and exendin-(9-39) an antagonist of the receptor. Diabetes 1993; 42:1678-1682.

362. Thorens B. Expression cloning of the pancreatic β cell receptor for the gluco-incretin hormone glucagon-like peptide 1. Proc Natl Acad Sci USA 1992; 89:8641-8645.

363. Chi L, Zhou W, Prikhozhan A et al. Cloning and characterization of the human GnRH receptor. Mol Cell Endocrinol 1993; 91:R1-R6.

364. Tsutsumi M, Zhou W, Millar RP et al. Cloning and functional expression of a mouse gonadotropin-releasing hormone receptor. Mol Endocrinol 1992; 6:1163-1169.

365. Reinhart J, Mertz LM, Catt KJ. Molecular cloning and expression of cDNA encoding the murine gonadotropin-releasing hormone receptor. J Biol Chem 1992; 267:21281-21284.

366. Illing N, Jacobs GFM, Becker II et al. Comparative sequence analysis and functional characterization of the cloned sheep gonadotropin-releasing hormone receptor reveal differences in primary structure and ligand specificity among mammalian receptors. Biochem Biophys Res Commun 1993; 196:745-751.

367. Eidne KA, Sellar RE, Couper G et al. Molecular cloning and characterisation of the rat pituitary gonadotropin-releasing hormone (GnRH) receptor. Mol Cell Endocrinol 1992; 90:R5-R9.

368. Chhajlani V, Wikberg JES. Molecular cloning and expression of the human melanocyte stimulating hormone receptor cDNA. FEBS Lett 1992; 309:417-420.

369. Chhajlani V, Muceniece R, Wikberg JES. Molecular cloning of a novel human melanocortin receptor. Biochem Biophys Res Commun 1993; 195:866-873.

370. Gantz I, Konda Y, Tashiro T et al. Molecular cloning of a novel melanocortin receptor. J Biol Chem 1993; 268:8246-8250.

371. Aharony D, Little J, Thomas C et al. Isolation and pharmacological characterization of a hamster urinary bladder neurokinin A receptor cDNA. Mol Pharmacol 1993; 45:9-19.

372. Li X-J, Wu Y-N, North A et al. Cloning, functional expression, and developmental regulation of a neuropeptide Y receptor from *Drosophila melanogaster*. J Biol Chem 1992;

267:9-12.

373. Wang J-B, Johnson PS, Persico AM et al. Human μ opiate receptor. cDNA and genomic clones, pharmacologic characterization and chromosomal assignment. FEBS Lett 1994; 338:217-222.

374. Schneider H, Feyen JHM, Seuwen K et al. Cloning and functional expression of a human parathyroid hormone receptor. Eur J Pharmacol 1993; 246:149-155.

375. Jüppner H, Abou-Samra A-B, Freeman M et al. A G protein-linked receptor for parathyroid hormone and parathyroid hormone-related peptide. Science 1991; 254:1024-1026.

376. Pausova Z, Bourdon J, Clayton D et al. Cloning of a parathyroid hormone/parathyroid hormone-related peptide receptor (PTHR) cDNA from a rat osteosarcoma (UMR 106) cell line: chromosomal assignment of the gene in the human, mouse, and rat genomes. Genomics 1994; 20:20-26.

377. Matsumoto K, Yokogoshi Y, Fujinaka Y et al. Molecular cloning and sequencing of porcine somatostatin receptor 2⁺. Biochem Biophys Res Commun 1994; 199:298-305.

378. Rasmussen UB, Vouret-Craviari V, Jallat S et al. cDNA cloning and expression of a hamster α-thrombin receptor coupled to $Ca^{2+}$ mobilization. FEBS Lett 1991; 288:123-128.

379. Vu T-K, Hung DT, Wheaton VI et al. Molecular cloning of a functional thrombin receptor reveals a novel proteolytic mechanism of receptor activation. Cell 1991; 64:1057-1068.

380. Duthie SM, Taylor PL, Anderson L et al. Cloning and functional characterisation of the human TRH receptor. Mol Cell Endocrinol 1993; 95:R11-R15.

381. Matre V, Karlsen HE, Wright MS et al. Molecular cloning of a functional human thyrotropin-releasing hormone receptor. Biochem Biophys Res Commun 1993; 195:179-185.

382. Yamada M, Monden T, Satoh T et al. Pituitary adenomas of patients with acromegaly express thyrotropin-releasing hormone messenger RNA: cloning and functional expression of the human thyrotropin-releasing hormone receptor gene. Biochem Biophys

Res Commun 1993; 195:737-745.

383. Straub RE, Frech GC, Joho RH et al. Expression cloning of a cDNA encoding the mouse pituitary thyrotropin-releasing hormone receptor. Proc Natl Acad Sci USA 1990; 87:9514-9518.

384. Zhao D, Yang J, Jones KE et al. Molecular cloning of a complementary deoxyribonucleic acid encoding the thyrotropin-releasing hormone receptor and regulation of its messenger ribonucleic acid in rat GH cells. Endocrinology 1992; 130:3529-3536.

385. de la Peña P, Delgado LM, del Camino D et al. Cloning and expression of the thyrotropin-releasing hormone receptor from GH₃ rat anterior pituitary cells. Biochem J 1992; 284:891-899.

386. Sellar RE, Taylor PL, Lamb RF et al. Functional expression and molecular characterization of the thyrotropin-releasing hormone receptor from the rat anterior pituitary gland. J Mol Endocrinol 1993; 10:199-206.

387. Satoh T, Feng P, Wilber JF. A truncated form of the thyrotropin-releasing hormone receptor is expressed in the rat central nervous system as well as in the pituitary gland. Mol Brain Res 1993; 20:353-356.

388. Thibonnier M, Auzan C, Madhun Z et al. Molecular cloning, sequencing, and functional expression of a cDNA encoding the human $V_{1a}$ vasopressin receptor. J Biol Chem 1994; 269:3304-3310.

389. Mahlmann S, Meyerhof W, Hausmann H et al. Structure, function, and phylogeny of [Arg⁸]vasotocin receptors from teleost fish and toad. Proc Natl Acad Sci USA 1994; 91:1342-1345.

390. Svensson SPS, Bailey TJ, Pepperl DJ et al. Cloning and expression of a fish $α_2$-adrenoceptor. Br J Pharmacol 1993; 110:54-60.

391. Jasper JR, Link RE, Chruscinski AJ et al. Primary structure of the mouse $β_1$-adrenergic receptor gene. Biochem Biophys Acta 1993; 1178:307-309.

392. Cohen JA, Baggott LA, Romano C et al. Characterization of a mouse $β_1$-adrenergic receptor genomic clone. DNA Cell Biol 1993; 12:537-547.

393. Yarden Y, Rodriguez H, Wong SK-F et al. The avian β-adrenergic receptor: primary

structure and membrane topology. Proc Natl Acad Sci USA 1986; 83:6795-6799.

394. Frail DE, Manelli AM, Witte DG et al. Cloning and characterization of a truncated dopamine D1 receptor from goldfish retina: stimulation of cyclic AMP production and calcium mobilization. Mol Pharmacol 1993; 44:1113-1118.

395. Martens GJM, Molhuizen HOF, Gröneveld D et al. Cloning and sequence analysis of brain cDNA encoding a *Xenopus* D₂ dopamine receptor. FEBS Lett 1991; 281:85-89.

396. De Backer MD, Gommeren W, Moereels H et al. Genomic cloning, heterologous expression and pharmacological characterization of a human histamine H1 receptor. Biochem Biophys Res Commun 1993; 197:1601-1608.

397. Tietje KM, Nathanson NM. Embryonic chick heart expresses multiple muscarinic acetylcholine receptor subtypes. Isolation and characterization of a gene encoding a novel m₂ muscarinic acetylcholine receptor with high affinity for pirenzepine. J Biol Chem 1991; 266:17382-17387.

398. Tietje KM, Goldman PS, Nathanson NM. Cloning and functional analysis of a gene encoding a novel muscarinic acetylcholine receptor expressed in chick heart and brain. J Biol Chem 1990; 265:2828-2834.

399. Onai T, FitzGerald MG, Arakawa S et al. Cloning, sequence analysis and chromosome localization of a *Drosophila* muscarinic acetylcholine receptor. FEBS Lett 1989; 255:219-225.

400. Shapiro RA, Wakimoto BT, Subers EM et al. Characterization and functional expression in mammalian cells of genomic and cDNA clones encoding a *Drosophila* muscarinic acetylcholine receptor. Proc Natl Acad Sci USA 1989; 86:9039-9043.

401. Arakawa S, Gocayne JD, McCombie WR et al. Cloning, localization, and permanent expression of a Drosophila octopamine receptor. Neuron 1990; 2:343-354.

402. Saudou F, Amlaiky N, Plassat J-L et al. Cloning and characterization of a *Drosophila* tyramine receptor. EMBO J 1990; 9:3611-3617.

403. Cerutis DR, Hass NA, Iversen LJ et al. The cloning and expression of an OK cell cDNA encoding a 5-hydroxytryptamine₁ᵦ receptor. Mol Pharmacol 1993; 45:20-28.

404. Van Obberghen-Schilling E, Vouret-Craviari V, Haslam RJ et al. Cloning, functional expression and role in cell growth regulation of a hamster 5-HT2 receptor subtype. Mol Endocrinol 1991; 5:881-889.

405. Yang W, Chen K, Lan NC et al. Gene structure and expression of the mouse 5-HT2 receptor. J Neurosci Res 1992; 33:196-204.

406. Yu L, Nguyen H, Le H et al. The mouse 5-HT₁c receptor contains eight hydrophobic domains and is X-linked. Mol Brain Res 1991; 11:143-149.

407. Witz P, Amlaiky N, Plassat J-L et al. Cloning and characterization of a *Drosophila* serotonin receptor that activates adenylate cyclase. Proc Natl Acad Sci USA 1990; 87:8940-8944.

408. Saudou F, Boschert U, Amlaiky N et al. A family of *Drosophila* serotonin receptors with distinct intracellular signalling properties and expression patterns. EMBO J 1992; 11:7-17.

409. Sugamori KS, Sunahara RK, Guan H-C et al. Serotonin receptor cDNA cloned from *Lymnaea stagnalis*. Proc Natl Acad Sci USA 1993; 90:11-15.

410. Pollak MR, Brown EM, Chou Y-HW et al. Mutations in the human $Ca^{2+}$-sensing receptor gene cause familial hypocalciuric hypercalcemia and neonatal severe hyperparathyroidism. Cell 1993; 75:1297-1303.

411. Kunz D, Gerard NP, Gerard C. The human leukocyte platelet-activating factor receptor. cDNA cloning, cell surface expression, and construction of a novel epitope-bearing analog. J Biol Chem 1992; 267:9101-9106.

412. Ye RD, Prossnitz ER, Zou A et al. Characterization of a human cDNA that encodes a functional receptor for platelet activating factor. Biochem Biophys Res Commun 1991; 180:105-111.

413. Katsuyama M, Sugimoto Y, Namba T et al. Cloning and expression of a cDNA for the human prostacyclin receptor. FEBS Lett 1994; 344:74-78.

414. Funk CD, Furci L, FitzGerald GA et al. Cloning and expression of a cDNA for the human prostaglandin E receptor EP₁ subtype. J Biol Chem 1993; 268:26767-26772.

415. Kunapuli SP, Mao GF, Bastepe M et al. Cloning and expression of a prostaglandin E receptor EP₃ subtype from human erythroleukaemia cells. Biochem J 1994; 298:263-267.

416. Adam M, Boie Y, Rushmore TH et al. Cloning and expression of three isoforms of the human EP₃ prostanoid receptor. FEBS Lett 1994; 338:170-174.

417. Takeuchi K, Abe T, Takahashi N et al. Molecular cloning and intrarenal localization of rat prostaglandin E₂ receptor EP₃ subtype. Biochem Biophys Res Commun 1993; 194:885-891.

418. Takeuchi K, Takahashi N, Abe T et al. Two isoforms of the rat kidney EP₃ receptor derived by alternative RNA splicing: intrarenal expression co-localization. Biochem Biophys Res Commun 1994; 199:834-840.

419. Sakamoto K, Ezashi T, Miwa K et al. Molecular cloning and expression of a cDNA of the bovine protaglandin F₂α receptor. J Biol Chem 1994; 269:3881-3886.

420. Abramovitz M, Boie Y, Nguyen T et al. Cloning and expression of a cDNA for the human prostanoid FP receptor. J Biol Chem 1994; 269:2632-2636.

421. Sugimoto Y, Hasumoto K, Namba T et al. Cloning and expression of a cDNA for mouse prostaglandin F receptor. J Biol Chem 1994; 269:1356-1360.

422. Namba T, Sugimoto Y, Hirata M et al. Mouse thromboxane A₂ receptor: cDNA cloning, expression and Northern blot analysis. Biochem Biophys Res Commun 1992; 184:1197-1203.

423. Saxe CL, Johnson R, Devreotes PN et al. Multiple genes for cell surface cAMP receptors in *Dictyostelium discoideum*. Devel Genet 1991; 12:6-13.

424. Parr CE, Sullivan DM, Paradiso AM et al. Cloning and expression of a human P₂ᵤ nucleotide receptor, a target for cystic fibrosis pharmacotherapy. Proc Natl Acad Sci USA 1994; 91:3275-3279.

425. Li X-J, Wolfgang W, Wu Y-N et al. Cloning, heterologous expression and developmental regulation of a *Drosophila* receptor for tachykinin-like peptides. EMBO J 1991; 10:3221-3229.

426. Monnier D, Colas J-F, Rosay P et al. NKD, a developmentally regulated tachykinin receptor in *Drosophila*. J Biol Chem 1992; 267:1298-1302.

427. Hagen DC, McCaffrey G, Sprague GF. Evidence the yeast *STE3* gene encodes a receptor for the peptide pheromone a factor: gene sequence and implications for the structure of the presumed receptor. Proc Natl Acad Sci USA 1986; 83:1418-1422.

428. Kaplan MH, Smith DI, Sundick RS. Identification of a G protein coupled receptor induced in activated T cells. J Immunol 1993; 151:628-636.

429. Njuki F, Nicholl CG, Howard A et al. A new calcitonin-receptor-like sequence in rat pulmonary blood vessels. Clin Sci 1993; 85:385-388.

430. Okazaki H, Ishizaka N, Sakurai T et al. Molecular cloning of a novel putative G protein-coupled receptor expressed in the cardiovascular system. Biochem Biophys Res Commun 1993; 190:1104-1109.

431. Dobner T, Wolf I, Emrich T et al. Differentiation-specific expression of a novel G protein-coupled receptor from Burkitt's lymphoma. Eur J Immunol 1992; 22:2795-2799.

432. Birkenbach M, Josefsen K, Yalamanchili R et al. Epstein-Barr virus-induced genes: first lymphocyte-specific G protein-coupled peptide receptors. J Virol 1993; 67:2209-2220.

433. Hla T, Maciag T. An abundant transcript induced in differentiating human endothelial cells encodes a polypeptide with structural similaritites to G-protein-coupled receptors. J Biol Chem 1990; 265:9308-9313.

434. Perez HD, Holmes R, Kelly E et al. Cloning of a cDNA encoding a receptor related to the formyl peptide receptor of human neutrophils. Gene 1992; 118:303-304.

435. Ye RD, Cavanagh SL, Quehenberger O et al. Isolation of a cDNA that encodes a novel granulocyte N-formyl peptide receptor. Biochem Biophys Res Commun 1992; 184:582-589.

436. Nothacker H-P, Grimmelikhuijzen CJP. Molecular cloning of a novel, putative G protein-coupled receptor from sea anemones structurally related to members of the FSH, TSH, LH/CG receptor family from

mammals. Biochem Biophys Res Commun 1993; 197:1062-1069.

437. Harrigan MT, Campbell NF, Bourgeois S. Identification of a gene induced by glucocorticoids in murine T-cells: a potential G protein-coupled receptor. Mol Endocrinol 1991; 5:1331-1338.

437a. Isimaa TP, Kiefer J, Liu ML et al. Isolation and chromosomal localization of a novel human G protein-coupled receptor expressed predominantly in the central nervous system. Genomics 1994; in press.

438. Saeki Y, Ueno S, Mizuno R et al. Molecular cloning of a novel putative G protein-coupled receptor (GPCR21) which is expressed predominantly in mouse central nervous system. FEBS Lett 1993; 336:317-322.

439. Eidne KA, Zabavnik J, Peters T et al. Cloning, sequencing and tissue distribution of a candidate G protein-coupled receptor from rat pituitary. FEBS Lett 1991; 292:243-248.

440. Kouba M, Vanetti M, Wang X et al. Cloning of a novel putative G-protein-coupled receptor (NLR) which is expressed in neuronal and lymphatic tissue. FEBS Lett 1993; 321:173-178.

441. Sreedharan SP, Robichon A, Peterson KE et al. Cloning and expression of the human vasoactive intestinal peptide receptor. Proc Natl Acad Sci USA 1991; 88:4986-4990.

442. Cook JS, Wolsing DH, Lameh J et al. Characterization of the RDC1 gene which encodes the canine homolog of a proposed human VIP receptor. Expression does not correlate with an increase in VIP binding sites. FEBS Lett 1992; 300:149-152.

443. Fukuda K, Kato S, Mori K et al. cDNA cloning and regional distribution of a novel member of the opioid receptor family. FEBS Lett 1994; 343:42-46.

444. Chee MS, Satchwell SC, Preddie E et al. Human cytomegalovirus encodes three G protein-coupled receptor homologues. Nature 1990; 344:774-777.

445. Nicholas J, Cameron KR, Honess RW. Herpesvirus saimiri encodes homologues of G protein-coupled receptors and cyclins. Nature 1992; 355:362-365.

446. Massung RF, Jayarama V, Moyer RW. DNA sequence analysis of conserved and unique regions of swinepox virus: identification of genetic elements supporting phenotypic observations including a novel G protein-coupled receptor. Virology 1993; 197:511-528.

447. Wilkie TM, Chen Y, Gilbert DJ et al. Identification, chromosomal location, and genome organization of mammalian G-protein-coupled receptors. Genomics 1993; 18:175-184.

448. Gorbulev V, Akhundova A, Grzeschik K-H et al. Organization and chromosomal localization of the gene for the human bombesin receptor subtype expressed in pregnant uterus. FEBS Lett 1994; 340:260-264.

449. Gerard NP, Bao L, Xiao-Ping H et al. Human chemotaxis receptor genes cluster at $19_{q13-3-13.4}$. Characterization of the human C5a receptor gene. Biochemistry 1993; 32:1243-1250.

450. Stoffel M, Espinosa R, Le Beau MM et al. Human glucagon-like peptide receptor gene. Localization to chromosome band 6p21 by fluorescence in situ hybridization and linkage of a highly polymorphic simple tandem repeat DNA polymorphism to other markers on chromosome 6. Diabetes 1993; 42:1215-1218.

451. Morris SW, Nelson N, Valentine MB et al. Assignment of the genes encoding human interleukin-8 receptor types 1 and 2 and an interleukin-8 receptor pseudogene to chromosome 2q35. Genomics 1992; 14:685-691.

452. Gao J-L, Kuhns DB, Tiffany HL et al. Structure and functional expression of the human macrophage inflammatory protein $1\alpha$/RANTES receptor. J Exp Med 1993; 177:1421-1427.

453. Vamvakopoulos NC, Rojas K, Overhauser J et al. Mapping the human melanocortin 2 receptor (adrenocorticotropic hormone receptor; ACTHR) gene (MC2R) to the small arm of chromosome 18 (18p11.21-pter). Genomics 1993; 18:454-455.

454. Uhl GR, Childers S, Pasternak G. An opiate-receptor gene family reunion. Trends Neurosci 1994; 17:89-93.

455. Befort K, Mattéi M-G, Roeckel N et al. Chromosomal localization of the δ opioid receptor gene to human 1p34.3-p36.1 and

mouse 4D bands by *in situ* hybridization. Genomics 1994; 20:143-145.

456. Nahmias C, Blin N, Elalouf J-M et al. Molecular characterization of the mouse β₃-adrenergic receptor: relationship with the atypical receptor of adipocytes. EMBO J 1991; 10:3721-3727.

457. Grandy DK, Civelli O. G-protein-coupled receptors: the new dopamine receptor subtypes. Curr Opin Cell Biol 1992; 2:275-281.

458. Schwengel DA, Nouri N, Meyers DA et al. Linkage mapping of the human thromboxane $A_2$ receptor (TBXA2R) to chromosome 19p13.3 using transcribed 3' untranslated DNA sequence polymorphisms. Genomics 1993; 18:212-215.

459. Fitzgibbon J, Appukuttan B, Gayther S et al. Localization of the human blue cone pigment gene to chromosome band 7q31.3-32. Human Genetics 1994; 93:79-80.

460. Nathans J, Thomas D, Hogness DS. Molecular genetics of human color vision: the genes encoding blue, green and red pigments. Science 1986; 232:193-202.

461. Krause JE, Bu J-Y, Takeda Y et al. Structure, expression and second messenger-mediated regulation of the human and rat substance P receptors and their genes. Regul Pept 1993; 46:59-66.

462. Gerard NP, Eddy RL, Shows TB et al. The human neurokinin A (substance K) receptor. Molecular cloning of the gene, chromosome localization, and isolation of cDNA from tracheal and gastric tissues. J Biol Chem 1990; 265:20455-20462.

463. Buell G, Schulz MF, Arkinstall SJ et al. Molecular characterisation, expression and localisation of human neurokinin-3 receptor. FEBS Lett 1992; 299:90-95.

464. Seibold A, Brabet P, Rosenthal W et al. Structure and chromosomal localization of the human antidiuretic hormone receptor gene. Am J Hum Genet 1992; 51:1078-1083.

465. Bao B, Kunoa G. Isolation and characterization of the gene encoding the rat α₁ᵦ adrenergic receptor. Gene 1993; 131:243-247.

466. Granneman JG, Lahners KN, Chaudry A. Characterization of the human β₃-adrenergic receptor gene. Mol Pharmacol 1993; 44:264-270.

467. Van Spronsen A, Nahmias C, Krief S et al. The promoter and intron/exon structure of the human and mouse β₃-adrenergic-receptor genes. Eur J Biochem 1993; 213:1117-1124.

468. Bensaid M, Kaghad M, Rodriguez M et al. The rat β₃-adrenergic receptor gene contains an intron. FEBS Lett 1993; 318:223-226.

# CHAPTER 2

# SIGNALING THROUGH G PROTEIN-COUPLED RECEPTORS

G protein-coupled receptors are now defined in molecular terms and rightly so. Their most salient characteristic is the possession of a number of structural motifs the most important of which is the hydrophobicity profile diagnostic of seven membrane-spanning domains. Accordingly some investigators have advanced the terms serpentine or heptahelical as alternative, and more structurally descriptive, names for this class of receptor. These have not caught on and probably won't. Although undoubtedly clumsy, the term G protein-coupled receptor is of both historical and functional significance. It encapsulates what is still the most relevant characteristic of this class of receptor: its ability to couple to G proteins.

## G PROTEINS

G proteins comprise part of a superfamily of proteins which bind and hydrolyse GTP. In addition to G proteins the larger group includes: the elongation factor Ef-Tu, involved in protein synthesis; small monomeric GTPases such as ras, which convey signals originating from receptor tyrosine kinases; and larger cytoskeleton-associated proteins such as dynamin.[1,2] All members of this superfamily switch between "on" and "off" states corresponding to the binding of GTP and its subsequent hydrolysis. In the case of G protein-coupled receptors this switching mechanism conveys information from an occupied receptor to some sort of effector system. The latter could be either an enzyme, usually involved in regulating the levels of certain intracellular messengers, or an ion channel. The G proteins are heterotrimeric. Binding and hydrolysis of GTP occurs on the α subunit, and it is usually this molecule which determines the nature of the effector response.[1,3-5] Until very recently the βγ subunits were considered to be functionally inert, important only for the localization of the holoenzyme to the cytoplasmic face of the plasma membrane, and for quenching the signal derived from the α-subunit. This is now known to be an over simplification (see below).

## 1. THE G PROTEIN CYCLE

The essential feature of G protein coupling is the mechanism whereby binding and hydrolysis of GTP leads to conformational changes in the αβγ complex, thereby promoting activation of the effector system.[1,3-5] This process is cyclic in nature (Fig. 2.1). In the inactive state the α-subunits are bound to GDP and are functionally restricted by association with the βγ complex. Release of GDP under

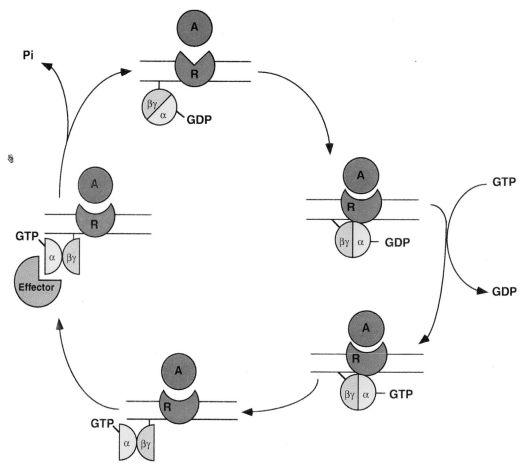

*Fig. 2.1. The G protein cycle. Binding of an agonist (A) to a G protein-coupled receptor (R) results in the exchange of GDP for GTP on the α-subunit of the G protein. This leads to the functional dissociation of the G protein heterotrimer and a reduction in the affinity of the receptor for its bound ligand. The activated α-subunit is free to interact with effector molecules until its bound GTP is hydrolyzed to GDP. This promotes the functional reassociation of the heterotrimer which is ready once more to interact with an occupied receptor if one is available.*

these conditions is very slow with a half-life in the order of seconds being common. Occupation of the receptor by its cognate ligand leads to a conformational change resulting in increased affinity of the receptor for the G protein holoenzyme. This in turn allows the rapid release of GDP from its binding site on the α subunit. Under experimental conditions, in the absence of guanine nucleotides, the binding site will remain empty and a stable complex is maintained between the G protein and the receptor which is bound tightly by its ligand. However, under physiological conditions, the GDP is immediately replaced

by GTP whose concentration probably exceeds that of GDP by several fold. The exchange of the guanine nucleotides leads to a reduction in the affinity of the α subunit for the βγ complex and functional dissociation of the heterotrimer. This has several major consequences. Firstly, the interaction between the receptor and G protein complex is diminished, resulting in a decrease in the affinity of the receptor for its cognate ligand and the release of the latter. The diminished coupling between the receptor and G protein also ensures that the cycle is driven forward: otherwise the occupied receptor might catalyze the wasteful re-exchange of GTP for GDP. The second important consequence of the functional dissociation of the G protein heterotrimer is the activation of the effector system by the GTP-bound α subunit. This whole process is self-regulated by the GTPase activity which is inherent to the α subunit. Hydrolysis of bound GTP to GDP promotes reassociation of the heterotrimer and turns off the signal. Provided that agonist concentrations have been maintained, and that some form of receptor desensitization has not occurred, rebinding of a ligand molecule will initiate another turn of the cycle.

It is this cyclic nature that explains many of the functional features of G protein-coupled receptors that were well-described before either the receptors themselves were structurally defined, or the details of the G protein cycle fully elucidated.[3] These features were the increased GTPase activity of membrane fractions to which receptor agonists had been added;[6] the decreased binding affinity of receptors for their ligands when assayed in the presence of GTP;[7] and the ability of nonhydrolysable GTP analogs to reproduce the functional responses of the native ligand and to render those responses insensitive to receptor occupation.[8] The scheme depicted in Figure 2.1 also provided an explanation for mechanisms of action of various bacterial toxins such as cholera and pertussis toxins.[9,10] Included amongst the species of α-subunit which are ADP-ribosylated by these toxins are $\alpha_s$ and $\alpha_i$, respectively involved in the stimulation and inhibition of the effector system adenylyl cyclase. In the case of the covalent modification of $\alpha_s$ by cholera toxin, this inhibits GTP hydrolysis, resulting in tonic activation of adenylyl cyclase.[9] In the case of $G_i$, ADP-ribosylation prevents the receptor-mediated activation of the G protein that would otherwise result in the inhibition of adenylyl cyclase.[10]

## 2. ADVANTAGES OF THE G PROTEIN CYCLE

The above discussion describes the functional role of G proteins in coupling between discrete receptor and effector molecules. However the more fundamental question is why such a complex system evolved. What are the advantages of introducing such a middleman over a more direct coupling between receptor and effector? Or for that matter why shouldn't agonist binding and effector function be mediated by a single molecular species as they are for the ligand-gated channels and receptor tyrosine kinases? One answer is to suppose that G proteins might perform functions in addition to coupling. Along these lines it has been proposed that α subunits might be released into the cytoplasm as "programmable second messengers", exerting effects far removed from the receptor.[11] However, although G protein activation is clearly linked with a decrease in the affinity between the α and βγ units, there is little direct evidence to suggest that the complex dissociates completely under physiological conditions.[12] This is demonstrated by the finding that α subunits can be active even when chemically cross-linked to βγ subunits.[13] This is not to say that the sole function of G proteins is restricted to the plasma membrane. Some cell types contain a pertussis toxin sensitive G protein localized to the Golgi region and which appears to play a role in the regulation of membrane trafficking.[14,15] It is possible that heterotrimeric G proteins might also translocate to the nucleus and thereby regulate gene transcription.[16]

Despite these interesting possibilities there is no real necessity to look beyond the plasma membrane in establishing the *raison d'etre* for G proteins. In fact a closer look at their function reveals that G proteins convey advantages to receptor/effector coupling in terms of sensitivity, amplification and flexibility of response.[3,5,12,17] G protein-coupled receptors occupy a middle ground both in terms of affinity and the time required to respond to changes in agonist concentration, when compared to ligand-gated channels (low affinity, rapid off-response) or receptor tyrosine kinases (high affinity, slow off-response).[17] The inverse relationship between affinity and response time is no coincidence in the single component systems: receptor tyrosine kinases are activated by very low concentrations of their circulating agonists, but once bound, release those agonists only slowly. This is ideally suited to the mediation of effects on cell proliferation and differentiation triggered by low abundance growth factors. The low affinity ligand-gated channels on the other hand respond to massive localized increases in neurotransmitter concentration, but promptly release their ligand once that concentration drops. This allows the rapid on/off responses underlying the conduction of nervous impulses. In both these examples the compromise between sensitivity and response time is a direct consequence of the fact that only the occupied receptor is active in terms of effector response. By separating the receptor and effector and, most importantly, by incorporating an on/off switch between them, the G protein-coupled receptors have been able to overcome this problem. This is because (with the exception of rhodopsin) exchange of GDP for GTP on the α subunit of the G protein is associated, not only with activation of the effector, but also with the dissociation of the bound ligand from its receptor. Therefore the duration of effector activation is not dependent on the rate of dissociation of the ligand from the receptor but on the GTPase activity of the α subunit. Of course in the longer term it will also depend on whether the agonist concentration remains sufficiently high to initiate the next cycle of ligand binding/G protein activation. As a consequence the G protein-coupled receptors display sensitivity both in terms of relatively high affinity for their cognate ligands, and in their abilities to respond rapidly to changes in agonist concentration. They also allow sensitivity in terms of agonist summation where multiple small signals derived from different receptor types can converge on a single G protein species to generate a physiologically meaningful effector response.

A second advantage of the functional dissociation of receptor binding and effector activation is the possibility of amplification.[3,17,18] Whereas single molecules of inactive G protein and receptor probably interact for only a few milliseconds, and ligand dissociation from the low affinity state of the receptor occurs in the sub-second range, the α-GTP moiety has a half life in the order of seconds. This means that a given receptor might sequentially bind and release several molecules of its ligand, and thereby activate several G proteins, before the signaling capacity of the first α-GTP is quenched. Although in principle the relatively slow rate of GTP hydrolysis might allow the activated α subunit to interact with several molecules of the effector enzyme, in practice this is unlikely to occur because of the low abundance of effector molecules relative to G proteins. However a second level of amplification is attained because an effector molecule, once activated, is turned off even more slowly than the GTP-bound α subunit: a single molecule of the latter may therefore give rise to many hundred molecules of intracellular messenger.[3,18] Amplification can thus be viewed as an important feature of G protein-coupled receptors and indeed helps to explain their sensitivity to comparatively low agonist concentrations.

Flexibility is the third advantage in the utilization of G proteins as on/off switches and is manifest in many forms. There is evidence that a single form of α subunit can interact with different effectors, and that both α and βγ subunits might be active.[3-5,12,19,20] This allows a wide range of responses, including diversification of signals from a single receptor, and integration of signals from multiple receptor types. These aspects will be discussed in detail below. A fourth advantage is that it offers an additional point of regulation. Although receptor occupation appears to be the only controlling factor for GDP-GTP exchange, the GTPase reaction on some α subunits has recently been shown to be regulated. The best described examples are those where certain effector enzymes, namely phospholipase C and cGMP phosphodiesterase, stimulate the GTP hydrolysis of their activator G proteins.[21,22] However, the regulator can also be an additional component of the system, such as the protein phosducin, which inhibits the GTPase activity of $G_s$.[23]

## 3. STRUCTURAL FEATURES OF G PROTEINS

At present molecular entities corresponding to 20 different mammalian α subunits have been described.[4,5,24] The βγ complex will be discussed in more detail below. The α species mostly all possess molecular weights in the range of 40-45 kDa but can be further subdivided on the basis of their sequence homology, functional characteristics and tissue distributions (Table 2.1). There are four subgroups: the $G_s$ group all of which stimulate adenylyl cyclase; the $G_i$ group originally described on the basis of inhibition of adenylyl cyclase but now known to include transducin ($α_T$) as well; the $G_q$ group responsible for activation of phospholipase C; and the $G_{12}$ group whose role is obscure. The first two groups are by far the best characterized. An exception is $α_z$ which has no known function but which has the interesting property of being able to hydrolyze GTP at less than one hundredth the rate of most other α subunits.[25]

Several of the regions of the various α subunits that are associated with the binding and hydrolysis of GTP have been recently defined.[1] These are highly conserved, but not contiguous with one another on the polypeptide chain. Mutation of some of the important amino acid residues in these regions results in inhibition of the GTPase activity and constitutive activation. Some of these mutations are naturally occurring and underlie the development of certain human endocrine tumors.[26,27] In particular, these studies have highlighted the importance of an arginine residue, at position 201 of $α_s$, for the efficient hydrolysis of GTP. This is the very residue which is ADP-ribosylated by cholera toxin.[1] Another mutation, resulting in an N-terminally truncated $α_s$ subunit, is responsible

for the disease known as Albright's hereditary osteodystrophy.[28] A short region at the N-terminus appears important for the association with βγ subunits.[29] Also contained at the N-terminus are the sites for myristoylation of $α_i$ and $α_o$, modifications which are essential for the correct association of those subunits with the plasma membrane.[30] Fatty acylation may regulate additional functions, such as association with βγs,[31] since $α_T$ has been shown to be covalently modified by a number of different lipid species.[32] In contrast, membrane-localization of $α_s$ is dependent on an intact C-terminus.[33] This is also the region that is thought to be most important for interaction of all α subunits with their receptors. The site for ADP-ribosylation of $α_i$ by pertussis toxin also occurs at the C-terminus only four residues from the end of the polypeptide chain.[1,5,18]

Very recently the first crystal structure of a G protein α subunit, that of $α_T$, has been published.[34] This has revealed a three-dimensional core structure very reminiscent of those of the ras oncogene and the elongation factor EF-Tu, despite the very limited identity in primary amino acid sequence which exists between the three proteins. This core forms the binding pocket for the guanine nucleotides as well as the surface with which the receptor interacts. In addition there is a helical region comprising 113 amino acids which is attached to the core by a linker sequence containing the $α_T$ equivalent of the $α_s$ A201. This helical region has no homolog in either ras or EF-Tu, both of which rely on additional molecules to regulate guanine nucleotide exchange and GTP hydrolysis. The helical region therefore appears to serve as a lid holding GDP in its binding pocket until the signal to open is delivered following receptor occupation.[34,35] However there is evidence that the helical region also functions catalytically in hydrolyzing GTP.[35] The assignation of both functions to a single region, and its incorporation into the core molecule, suggests the workings of an elegant evolutionary adaptation.

*Table 2.1. Features of mammalian G-protein α subtypes*

| Family Subtype | Expression | Effector System | Cognate Receptors |
|---|---|---|---|
| **G$_s$** | | | |
| α$_s$ (4 splice variants) | Ubiquitous | ↑ Adenylyl Cyclase<br>↑ Ca$^{2+}$ Channels<br>↓ Na$^+$ Channels | β-adrenergic,<br>glucagon, etc. |
| α$_{olf}$ | Olfactory | ↑ Adenylyl Cyclase | Odorant |
| **Gi** | | | |
| α$_{i1}$ | Widespread | ↓ Adenylyl Cyclase? others | α$_2$-adrenergic, |
| α$_{i2}$ | Ubiquitous | ↓ Adenylyl Cyclase, others | m2 muscarinic, |
| α$_{i3}$ | Widespread | ↑ K$^+$ Channels, others | somatostatin,<br>neuropeptideY, |
| α$_o$ (2 splice variants) | Neuroendocrine | ↑ K$^+$ Channels,<br>↓ Ca$^{2+}$ Channels | galanin, etc |
| α$_g$ | Taste buds | Unknown | Gustatory |
| α$_{T1}$ | Retinal Rods | ↑ cGMP phophodiesterase | Rhodopsin |
| α$_{T2}$ | Retinal Cones | ↑ cGMP phophodiesterase | Color pigments |
| α$_z$ | Neuroendocrine,<br>Circulatory | Unknown | Unknown |
| **Gq** | | | |
| α$_q$ | Widespread | ↑ PLC | m1 muscarinic, |
| α$_{11}$ | Widespread | ↑ PLC | α1-adrenergic |
| α$_{14}$ | Widespread | ↑ PLC | Unknown |
| α$_{16}$ | Circulatory | ↑ PLC | Unknown |
| **G12** | | | |
| α$_{12}$ | Ubiquitous | Unknown | Unknown |
| α$_{13}$ | Ubiquitous | Unknown | Unknown |

## EFFECTOR SYSTEMS

### 1. STIMULATION OF ADENYLYL CYCLASE

Although the second messenger function of cAMP, and the existence of the enzyme responsible for its generation, adenylyl cyclase, had been known since the late 1950s,[36] it took the best part of another two decades to unravel completely the mechanism whereby receptor binding by a hormone (primary messenger) led to the increase in cAMP.[3] Important steps along the way included demonstrations that the receptor and adenylyl cyclase were separate entities;[37] that GTP was needed for activation;[38] and that a third component, a G protein, was required to couple the receptor to the effector.[39-40] At first the specificity of the G protein for guanosine, as opposed to other nucleotides was un-

clear, and they were referred to as N-proteins. Thus G$_s$ and G$_i$ were commonly called N$_s$ and N$_i$ prior to the mid 1980s. The receptor agonists glucagon and epinephrine, used for these early studies, thus defined the first G protein-coupled receptors years in advance of their isolation and cloning. Moreover the details of G protein function worked out initially for the adenylyl cyclase system have subsequently been shown to hold true for many other examples of receptors and effector systems. This has gone hand in hand with the extension of the second messenger paradigm and the recognition of the importance of reversible protein phosphorylation as an integral component of the process of cellular activation. These concepts will now be expanded in the context of the adenylyl cyclase system.

According to the original definition of

Sutherland a molecule must satisfy four postulates in order to be considered as a bona fide second messenger. First, its concentration must be altered in response to the primary stimulus. Second, it must activate the cellular response in a manner consistent with that of the primary stimulus. Third, mechanisms for removal of the second messenger and the consequent termination of the signal must be present in the cell. Fourth, and most importantly, a causal relationship between agonist addition, second messenger generation and physiological response must be established. This can be addressed by the use of specific inhibitors and by detailed analyses of the time courses of generation of the second messenger and physiological response. Although it is now well established that cAMP satisfies all of these criteria they should be kept in mind for assessing other molecules whose credentials as second messengers are sometimes overstated.

Adenylyl cyclase is the enzyme which catalyses the formation of cAMP from ATP. The mammalian enzyme is currently thought to comprise a family of eight isotypes which display differing tissue distributions and activation characteristics (Table 2.2).[41-42] They are most probably all activated by the GTP-bound form of $\alpha_s$ (or $\alpha_{olf}$), but differ in their sensitivities to other regulators such as $Ca^{2+}$ (either directly, or in the form of the $Ca^{2+}$-binding protein calmodulin) and in their susceptibility to be stimulated, inhibited or unaffected by βγ subunits. These characteristics will obviously play an important role in determining how signals from different receptors are integrated to produce the net cellular response. The mammalian enzymes have molecular weights in the range of 110-120 kDa and are integral membrane proteins. Although overall amino acid identity between the various isoforms is limited, they all possess a highly characteristic domain structure which is not dissimilar to that of some ion channels. They contain two highly hydrophobic domains each thought to comprise six membrane-spanning helices. There are also two cytoplasmic domains, one existing as a C-terminal tail, the other as a loop between the two hydrophobic regions. It is the cytoplasmic domains which show the greatest sequence homology, both to each other, and to other members of the adenylyl cyclase family.[41-42]

The other important enzyme in the cAMP story is that through which it actually transmits its signal: the cAMP-dependent protein kinase (PKA). This exists in the inactive form as a tetramer comprising two regulatory and two catalytic subunits. Two molecules of cAMP bind cooperatively to each of the regulatory subunits causing them to dissociate and thereby allowing the substrate-binding surface of the catalytic subunit to be unmasked.[43] The latter is now free to interact with specific substrate proteins which may be present at various cellular locations including the nucleus.[44]

## Table 2.2. Properties of adenylyl cyclase subtypes

| Subtype | Tissue Distribution | Regulators |
|---------|---------------------|------------|
| 1 | Brain | ↑$\alpha_s$ ↑$Ca^{2+}$ CaM ↓βγ |
| 2 | Neuronal, Lung | ↑$\alpha_s$ ↑βγ ↑PKC |
| 3 | Olfactory | ↑$\alpha_{olf}$ |
| 4 | Widespread | ↑$\alpha_s$ ↑βγ |
| 5 | Heart, Striatum, Others | ↑$\alpha_s$ |
| 6 | Heart, Striatum, Others | ↑$\alpha_s$ ↓$Ca^{2+}$ |
| 7 | Brain, Erythroleukemia Cells | ↑$\alpha_s$ ? |
| 8 | Brain, Liver | ↑$\alpha_s$ ? |

This interaction results in the catalyzed transfer of γ-phosphate groups from ATP to serine or threonine residues present in defined regions of the substrate proteins. This covalent modification brings about a conformational change in the target protein which might alter its catalytic properties or its ability to interact with its own set of targets.[43] Reversal of this process is brought about by a small group of phosphatases.[45]

Subtle variations in this overall process have become apparent with recognition of the existence of four types of regulatory chain (RIα, RIβ, RIIα, RIIβ), along with three catalytic chains (Cα, Cβ, Cγ).[46-49] The α chains of both subunits are widely expressed, whereas the β subtypes are relatively restricted to neural tissue, and Cγ is specific to the testis. Two predominant PKA holoenzymes, types I and II, are assembled from these subunits, comprising homodimers of RI and RII respectively, along with the C chain dimer that is specific to the particular cell type. Although the mechanism of activation by cAMP is broadly similar, there are differences in response depending on which R subunits are present.[47-48] Whereas RI is relatively diffuse in the cytoplasm, a large proportion of the inactive RII containing enzyme is tethered to the cytoskeleton. The type II subunits can also be autophosphorylated by the catalytic chains, which appears to inhibit reassembly of the inactive holoenzyme. Signaling through type II PKA might therefore be more prolonged than that through the type I enzyme. In addition, the type Iβ subunit is activated at lower concentrations of cAMP than the type Iα, suggesting that differential tissue distributions of these subunits might be accompanied by differential sensitivities to agonists working through adenylyl cyclase. At present no differences between the catalytic properties of the various C subunits have been found.

## 2. INHIBITION OF ADENYLYL CYCLASE

An important subfamily of the G protein-coupled receptors are those which are coupled to inhibition of adenylyl cyclase.[3] Once again the functional description of this system long preceded its molecular characterization. The G protein involved was distinct from $G_s$ as it was a substrate for pertussis, but not cholera, toxin. However it was soon found that a number of proteins in tissue extracts could be ADP-ribosylated by pertussis toxin. This was confirmed by molecular cloning experiments demonstrating the existence of three species of $G_i$, and two of $G_o$, all of which were pertussis toxin-sensitive.[3,24,50] It is only very recently that the weight of evidence has begun to point to $\alpha_{i2}$ as the entity chiefly responsible for conveying inhibition of adenylyl cyclase.[20]

Controversy raged for some years over the actual mechanism underlying this inhibitory effect.[18] Some workers contended that it was mediated by βγ subunits which, upon functional dissociation from $G_i$, became available to complex with the α subunit of $G_s$, thus rendering it inactive. Such a mass action effect would be possible because of the vast excess of $G_i$ over $G_s$ in virtually all tissues. An alternative theory was that $\alpha_i$ was the active moiety, binding directly to adenylyl cyclase and inhibiting its activity. It is the latter of two original hypotheses which has held up best in the light of recent work. The availability of pure GTP-bound $\alpha_i$ has led to the demonstration that it directly inhibits adenylyl cyclase, or at least the three out of seven isoforms that have been currently tested.[51] Interestingly, however, a role for $G_i$-derived βγ has also emerged, not by interaction with $\alpha_s$, but by direct binding with a limited number of adenylyl cyclase isoforms.[41] This will be discussed in detail below.

It has been apparent for more than a decade that inhibition of adenylyl cyclase activity constitutes only a fraction of the global effects triggered by occupation of the inhibitory G protein-coupled receptors.[52-54] They can also activate $K^+$ channels, thereby hyperpolarizing cells, and attenuate $Ca^{2+}$ influx through voltage-gated $Ca^{2+}$ channels; both effects are inhibitory in electrically excitable cells.[55-57] An additional, but poorly understood mechanism also appears to operate distally to the generation of all known intracellular messengers. This mechanism has been repeatedly described in secretory cells although its exact site is unclear.[58-59] It most probably corresponds to a blockade of one of the final steps of the exocytotic process, perhaps even to fusion of the secretory granule with the plasma membrane. Although the species of G protein involved is pertussis toxin-sensitive, its exact identity is unknown.[58-59]

## 3. Photoreceptor Signaling

Mammalian photoreceptor cells have evolved a highly specialized system for detecting changes in illumination.[60] The best characterized are the rod cells which selectively express the G protein-coupled receptor rhodopsin which contains within its binding pocket a chromophore, 11-*cis*-retinal, that detects photons and is thereby converted to the all-*trans* form. The G protein that couples to activated rhodopsin is called transducin ($G_T$) which exists in two forms, one each expressed exclusively in either rod or cone cells.[24] The effector is a phosphodiesterase enzyme which hydrolyses the cyclic nucleotide cGMP. This second messenger is produced by a guanylyl cyclase which, unlike adenylyl cyclase is not under the control of G protein-coupled receptors.[60] The cGMP phosphodiesterase enzyme is one of a family of five isoforms that show varying specificities for cAMP versus cGMP, and catalyze the formation of 5'-AMP and 5'-GMP, respectively.[61] The phosphodiesterases also display different tissue distributions and in their sensitivities to be stimulated by $Ca^{2+}$-calmodulin. The retinal enzyme is comprised of three types of polypeptide chain $\alpha$, $\beta$ and $\gamma$ with molecular weights of 88, 85 and 9 kDa, respectively. The holoenzyme consists of single copies of each of the $\alpha$ and $\beta$ subunits and a $\gamma$ subunit dimer. Interaction with transducin leads to removal of the $\gamma$ subunits thereby allowing the catalytic activity of the $\alpha\beta$ subunit to be expressed.[60-61] This also results in an increase in the GTPase

activity of $\alpha_T$, thereby autoregulating the signal and ensuring that not too much cGMP is hydrolysed.[22] In most cell types the immediate target of cGMP is the cGMP-dependent protein kinase. However, the main role of cGMP in photoreceptors cells is the maintenance in an open state of a cation-specific channel capable of conducting both $Na^+$ and $Ca^{2+}$. Activation of photoreceptors is therefore associated with a G protein-coupled reduction in cGMP levels, blockade of the cation-specific channel and a consequent hyperpolarization and decrease in the cytosolic free $Ca^{2+}$ concentration.[60] Although not of immediate relevance to any cell type other than photoreceptors, the rhodopsin system has been widely studied, and has yielded much general information on the functions and characteristics of G proteins in receptor-effector coupling. This is especially true for the process of desensitization as will be discussed below.

## 4. Phospholipase C

The realization over the last decade or so that phospholipase C (PLC) acts as an effector enzyme, in a manner analogous to adenylyl cyclase, has been a major factor contributing to the growing interest in G protein-coupled receptors. In some respects it is fortunate that the significance of PLC was not discovered too soon after adenylyl cyclase, since it is a much more complex system.[62-63] Firstly, not one but two second messengers are produced. Secondly, the further metabolism of both of these is complex and may give rise to other signal mediators. Thirdly, only one of the second messengers interacts directly with a kinase; the other releases $Ca^{2+}$ which is therefore sometimes described, rather clumsily, as a third messenger.

The substrate for PLC is phosphatidylinositol 4,5-bisphosphate ($PIP_2$) a minor lipid component of the plasma membrane.[62-63] Whereas ATP would be an important molecule, even if it wasn't the precursor for cAMP, $PIP_2$ appears to have been designed especially for its role in cellular activation. Even a possible function in anchoring cytoskeletal elements to the plasma membrane [64-65] is probably just another facet of this signaling role. Like all other phospholipids, PIP$_2$ contains a 3-carbon backbone derived from glycerol. The first two carbon atoms are linked to long chain fatty acyl moieties and the third is attached to a polar head group via a phosphate linkage (Fig. 2.2). In the case of PIP$_2$ this head group comprises a 6-carbon inositol ring, substituted on the 4 and 5 positions with phosphate groups. The latter are added sequentially to the precursor compound phosphatidylinositol (PI) forming phosphatidylinositol 4-monophosphate (PIP) and PIP$_2$, respectively (Fig. 2.3). PIP$_2$ itself represents only a few percent of the total phosphoinositide pool. Whereas all three act as substrates for PLC in vitro,[66-67] hard evidence for the utilization of anything other than PIP$_2$ in vivo has been obtained for only a very few cell types.[68-69] The critical reaction is therefore the hydrolytic cleavage of PIP$_2$ to form inositol 1,4,5-trisphosphate (IP$_3$), corresponding to the original polar head group, and diacyl-glycerol (DAG) comprising the glycerol backbone with its constituent fatty acid side chains (Fig. 2.2).

Within a few years of the demonstration of the second messenger function of IP$_3$ and DAG, evidence had also appeared to suggest that receptor occupation and PLC activity were coupled via a G protein. This was based on the use of GTP and its nonhydrolysable analogs in cell free systems either to inhibit receptor binding, or to synergise with ligands in the generation of PIP$_2$ breakdown products.[67,70-71] For nearly six years the G protein was unidentified and known by default as G$_p$. Then in early 1991 two groups demonstrated activation of PLC activity in reconstitution experiments using the purified $\alpha$ subunit of the G protein, G$_q$.[72-73] The latter had been discovered some years earlier by molecular cloning but had, until then, no known function.[74] G$_q$ and its close relative G$_{11}$ are pertussis toxin-insensitive. Yet activation of PLC in some cells, especially those of hemopoietic origin, was known to be abolished by the toxin. This discrepancy has only been resolved very recently with a better understanding of the function of $\beta\gamma$ subunits (see below).

PLC refers to any phosphodiesterase enzyme cleaving a phospholipid above the phosphate linkage to which the head group is attached (Fig. 2.2). The PLC enzymes which hydrolyze the phosphoinositides are therefore a subset of a much wider family. Accordingly they are referred to as phosphoinositidase(s) C by some workers. Despite its obvious advantage this nomenclature has not caught on, and generally the term "PI-specific" is used if any qualification of the generic PLC is needed. This should not be taken as indicating any preference

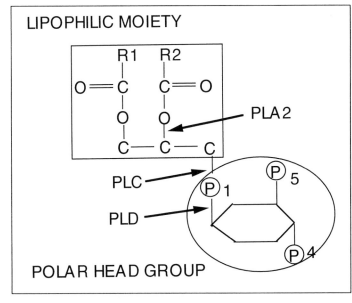

Fig. 2.2. *Structural representation of phosphatidylinositol 4,5-bisphosphate (PIP$_2$). The lipophilic and polar moieties of the molecule are displayed as are the sites at which the various phospholipases A$_2$, C and D are known to act. R1 is a long-chain fatty acyl group; R2 is often, but not exclusively, an arachidonyl group.*

*Fig. 2.3. The phosphoinositide cycle. Diacylglcerol (DAG) formed through the action of phospholipase C (PLC) on phosphatidylinositol 4,5-bisphosphate (PIP₂) is degraded by one of two routes: conversion to phosphatidic acid (PA) which predominates in most cells and is catalyzed by DAG lipase; or removal of arachidonic acid (ARA) by DAG lipase, with the con-comitant generation of lyso-*

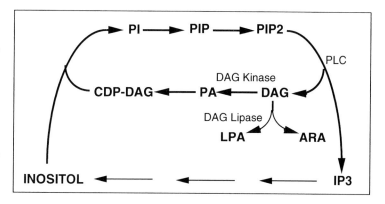

*phosphatidic acid (LPA). Inositol 1,4,5-trisphosphate (IP₃) is metabolized to inositol, which recombines with a CDP-derivative of DAG to form phosphatidylinositol (PI) a precursor of PIP₂.*

for PI over $PIP_2$, but as specific for phosphoinositides over other phospholipids. The PI-specific PLCs are grouped in three main families of which the PLCβ family are known to be G protein-linked, the PLCγs interact with receptor tyrosine kinases, and the PLCδ have no known activation mechanism.[66] Ironically the PLCα family is no longer believed to be a PLC at all. There is only limited homology between the three families, confined chiefly to two regions denoted X and Y. PLCβ is now known to comprise four isoforms.[66,75-77] PLCβ1 is widely expressed and has been shown to be the predominant isoform coupling to $G_{q/11}$.[66,73] PLCβ2 was cloned originally from hemopoietic cells and most probably corresponds to the pertussis toxin-sensitive activity associated with those cells and mediated by βγ subunits.[75,78-80] PLCβ3 appears capable of coupling to both. All of the members of the PLCβ family show a $Ca^{2+}$ requirement and act preferentially on $PIP_2$; at high $Ca^{2+}$ concentrations they can also hydrolyze PI independently of G protein involvement, but this is unlikely to be of any physiological significance.[66,75,79]

Of the two intracellular messengers formed by hydrolysis of $PIP_2$, only $IP_3$ is water soluble and capable of diffusing through the cytosol.[81] It exerts its effects by binding to a receptor most probably present on a subcompartment of the smooth endoplasmic reticulum.[81-82] This organelle accumulates $Ca^{2+}$ on the luminal

side by virtue of a high affinity, ATP-dependent $Ca^{2+}$ pump. Three major forms of $IP_3$ receptor, types I-III, have been characterized molecularly.[83-85] The type I receptor, at least in its predominant form, is neuronal, whereas an alternatively spliced form, as well as types II and III, are chiefly expressed in nonneuronal tissues.[83-86] All forms display substantial homology to the ryanodine receptor $Ca^{2+}$ release channel present in the T tubules of skeletal muscle. Like the ryanodine receptor, the functional unit of the $IP_3$ receptor is thought to be a homotetramer.[81] The four subunits form a central pore through which $Ca^{2+}$ passes down its concentration gradient from the lumen of the endoplasmic reticulum to the cytosol. Gating of this channel is triggered by $IP_3$ which binds co-operatively to a single site on the C-terminus of each of the subunits. The net consequence is a rise in the cytoplasmic free $Ca^{2+}$ concentration. This had long been recognized as a key event in many examples of cellular activation, but its link to receptor occupation had been obscure until the discovery of the function of $IP_3$ in late 1983.[87]

Many proteins within the cell are capable of binding $Ca^{2+}$ but probably the most important in terms of cellular signaling is calmodulin.[88] Depending on the extent of the rise in cytosolic free $Ca^{2+}$ calmodulin binds anywhere up to four $Ca^{2+}$ ions, each inducing a discrete conformational change in the protein. Calmodulin does not possess any intrinsic catalytic activity but acts rather as an adaptor protein interacting with a diverse, but defined, range of targets depending on the prevailing $Ca^{2+}$ concentration. Many of these targets are protein kinases, such as $Ca^{2+}$-calmodulin dependent protein kinase, or key metabolic enzymes such as phosphorylase.

Another target of $Ca^{2+}$-calmodulin is $IP_3$ kinase (Fig. 2.4) the enzyme which, at the expense of a molecule of ATP, phosphorylates $IP_3$ to inositol 1,3,4,5-tetrakisphosphate ($IP_4$).[89] Since $IP_3$ is also sequentially dephosphorylated, yielding eventually free inositol, it has been argued that this energetically unfavorable, and highly regulated, alternative pathway must serve a function in addition to removal of $IP_3$.[89] According to one conceptually elegant hypothesis, the additional function might be the generation of $IP_4$. The latter has been proposed as an intracellular messenger, promoting $Ca^{2+}$ influx from the

extracellular space and thereby prolonging the rise in cytosolic free $Ca^{2+}$ which would otherwise be restricted by the limited capacity of the intracellular stores.[89] However, while it is clear that $Ca^{2+}$ influx is an important consequence of the activation of PLC, the great majority of cells seem capable of controlling that influx without any obligatory role for $IP_4$. On the other hand there is growing evidence that $Ca^{2+}$ influx is regulated by the degree of depletion of the intracellular store.[90-91] Whether the store communicates directly with the plasma membrane, or whether it does so indirectly, through a mediator[92] or small monomeric G protein,[93-94] or whether indeed $IP_4$ plays any role at all in this process is currently unresolved. The nature of the channel is also unknown, although it clearly differs from the voltage-gated channels in terms of pharmacology and electrophysiological properties.[81]

An alternative explanation for the complexity of $IP_3$ metabolism is simply the

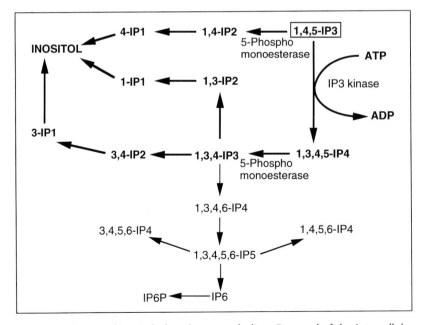

Fig. 2.4. Pathways of inositol phosphate metabolism. Removal of the intracellular messenger inositol 1,4,5-trisphosphate (1,4,5-IP₃) is controlled by two enzymes: a 5-phosphomonoesterase; and a 3-kinase. The former enzyme initiates a route which would be the simplest and most energetically favorable if degradation of IP3 were the only consideration. The alternative pathway gives rise to a host of metabolites. Possible explanations for this proliferation of metabolites are discussed in the text.

multiple specificity of a handful of inositol phosphomonoesterase enzymes.[89,95-96] This is apparent from a consideration of the relevant biochemical pathways (Fig. 2.4). The 5-phosphomonoesterase which dephosphorylates $IP_3$ also removes the phosphate group from the 5 position of $IP_4$. Similarly, a 1-phosphomonoesterase can also act on several substrates and so on. While this conserves the number of enzymes necessary for regeneration of free inositol, it does result in a proliferation of different intermediates. However this explanation does not clarify the role of $IP_3$ kinase, nor that of the other, more recently discovered, kinase enzymes that generate even more highly phosphorylated inositol derivatives (Fig. 2.4).[97] Unfortunately there is no way yet of knowing whether the effort and skill that has gone into elucidating these novel pathways and metabolites will be rewarded with the discovery of any physiologically relevant function. The simplest but least satisfying explanation would be the vestigial retention in mammals of enzymes important in lower organisms, such as birds, where $IP_6$ plays a role in the regulation of oxygen tension in erythrocytes.[89]

The other second messenger produced concomitantly with $IP_3$ is DAG. Being lipophilic this is retained within the plane of the plasma membrane, meaning that its targets must be either membrane-bound or capable of membrane association. The chief target is PKC. This was originally discovered as a proteolytically activated protein kinase, but subsequently shown to be regulated by DAG in a $Ca^{2+}$ and phospholipid dependent manner.[98-99] It is now known to comprise a family of at least 10 different isoforms (Table 2.3).[99,100-101] Two findings brought PKC into prominence in the early 1980s: the recognition of the importance of $PIP_2$ hydrolysis for cellular activation;[62] and the identification of PKC as the endogenous target for the tumor-promoting phorbol esters.[102] PKC was subsequently shown to be a cytosolic enzyme in nonstimulated cells but to become membrane associated following receptor occupation.[103] This translocation made physiological sense in terms of gaining access to DAG, and was shown to be due to the rise in the cytosolic free $Ca^{2+}$ concentration accompanying $IP_3$ generation.[104] Information on the structure of PKC, derived from the molecular cloning of the first family of isoforms, tended to support this scenario.[105,106] In addition to a catalytic, or kinase domain, similar to that of PKA and other serine/threonine kinases, PKC possessed a large regulatory domain containing regions of postulated importance for the binding of DAG, phospholipid and $Ca^{2+}$. These broad assignations have been

## Table 2.3. Properties of PKC isoforms

| Family<br>Isoform | Calculated<br>Size (kDa) | Expression | Activators |
|---|---|---|---|
| **Classical** | | | |
| α | 77 | Ubiquitous | DAG (Phospholipid |
| β (I and II) | 77 | Widespread | and $Ca^{2+}$-dependent) |
| γ | 78 | Neuronal | |
| **Novel** | | | |
| δ | 78 | Widespread | DAG (Phospholipid |
| ε | 83 | Widespread | dependent) |
| ν | 78 | Skin, Lung | |
| θ | 82 | Skeletal Muscle | |
| **Atypical** | | | |
| ζ | 68 | Ubiquitous | Unknown |
| ι | 67 | Lung, others | |

largely confirmed by subsequent mutagenesis studies.[48,107]

A still unresolved question is how PKC recognizes DAG generated in response to receptor occupation as opposed to the excess already present as a constituent of the membrane bi-layer. It is probable that a protein/protein interaction provides the required specificity for targeting PKC. Candidates would be the recently described "receptors for activated C-kinase", proteins associated with the subplasmalemmal cytoskeleton which appear to bind only the activated PKC.[108] Another largely unresolved problem with the translocation model is the mechanism whereby cytosolic proteins become phosphorylated. Recent work addressing this question provides some evidence that, in fact, translocation of PKC may neither be necessary nor sufficient for activation.[109-110] This is an especially important issue for the novel family of PKC isoforms (Table 2.3). These are neither $Ca^{2+}$-dependent nor as clearly defined as the classical PKCs in terms of spatial distribution.[99,100,110] The very recently described atypical PKC family are even more problematic (Table 2.3). In addition to lacking a $Ca^{2+}$-binding domain, they possess only a single copy of the cysteine-rich, phospholipid-binding domain which exists as a tandem repeat in the other PKC families.[99,101] The atypical PKCs are activated by neither DAG analogs nor phorbol esters, and translocate only in a very few cell types.[99,110] It is thus possible that their signaling potential is linked to systems other than G protein-coupled receptors.

The major route of DAG removal in most cells is via phosphorylation to phosphatidic acid, catalyzed by the enzyme DAG kinase (Fig. 2.3).[111] As such phosphatidic acid is an intermediate in the cyclic reconversion of DAG back to PI, and hence $PIP_2$. It has been perennially touted as an intracellular messenger, but with little in the form of hard evidence to recommend it. A better candidate as a signaling molecule is arachidonic acid which can be formed via DAG lipase in the other major route of DAG degradation (Fig. 2.3).

Although little studied, the recent availability of specific inhibitors of this pathway has led to suggestions that it might represent an important means of diversifying signaling potential in some cells.[112-113]

By analogy with the adenylyl cyclase system it might be predicted that receptor-coupled inhibition of PLC activity would also be well-documented. This is not the case. Although some receptors do couple to a functional inhibition of PLC, this is almost certainly not a direct effect, but due rather to an attenuation of $Ca^{2+}$ influx and the consequent abolition of the potentiating effect of raised cytosolic free $Ca^{2+}$ on PLC activity.[54,114] In cell-free systems, where the evidence is strongest for a direct involvement of a G protein, the inhibition of PLC activity is rather small.[115,116] This was especially true of the only study in which a G protein $\alpha$ subunit $(\alpha_i)$ has been clearly implicated.[116] It is perhaps more likely that the slight inhibition of PLC activity that has been described in conjunction with adenosine receptors, and which appears independent of alterations in $Ca^{2+}$ influx and adenylyl cyclase, might therefore be due to some cross-talk response mediated by $\beta\gamma$ subunits.[114]

## 5. OTHER EFFECTOR SYSTEMS

Adenylyl cyclase and PLC are without doubt the most important enzymes coupled to G protein-coupled receptors. However, other enzyme effector systems have also been described. The longest studied has been $PLA_2$, the enzyme which releases arachidonic acid from the 2 position of phospholipids (Fig. 2.2). In the late 1980s it was shown that this process could be directly activated by GTP analogs and in a manner apparently independently of $PIP_2$ hydrolysis.[117] This latter feature distinguished the mechanism for the much more common stimulation of $PLA_2$ that is secondary to rises in either cytosolic free $Ca^{2+}$ or regulators of PKC.[118] Despite widespread study, the direct mechanism appears confined to a very limited number of cell types and receptor species. Although pertussis toxin-sensitive, the exact identity of the

G protein involved has not been determined; nor has that of the $PLA_2$ isoform, although the recently cloned, cytosolic form of the enzyme would be an obvious candidate.[118] Arachidonic acid is the precursor for a host of biologically active molecules which are released from the cell and act in an autocrine fashion. Although sometimes referred to as a second messenger, arachidonic acid itself has not yet fully satisfied all of the relevant criteria especially in terms of sufficiency of reproducing the effects of the agonist. Nor are its intracellular targets known with any certainty. Arachidonic acid acts in vitro both to raise cytosolic free $Ca^{2+}$ by inhibition of $Ca^{2+}$ uptake into intracellular stores,[119] and to activate certain PKC isoforms.[99] The physiological significance of either of these actions is currently unknown.

In addition to PLC there is evidence that phospholipase D (PLD) plays an important role in some instances of cellular activation (Fig. 2.2).[120-121] The substrate is generally phosphatidylcholine and the resultant products are choline and phosphatidic acid. Conversion of the latter to DAG is generally thought to underlie the signaling function of the PLD pathway. Although there is some evidence for direct G protein involvement in the activation of PLD, a PKC-dependent pathway is more widespread.[99] The identity of the G protein is unknown and the PLD enzymes involved in signaling are entirely uncharacterized.

An exciting recent addition to the ranks of G protein-coupled effector enzymes are the tyrosine phosphatases. As the name suggests these enzymes remove phosphate groups from the substrates of tyrosine kinases and as such may be key players in integrating signals from receptor tyrosine kinases and G protein-coupled receptors. The evidence is based on a somatostatin-stimulated tyrosine phosphatase activity in the membranes of some cell types that was potentiated by GTP but inhibited by GDP analogs and prior treatment with pertussis toxin.[122] Although potentially exciting, these results have yet to be confirmed or extended to other experimental systems. On the other hand there have been reports implicating receptors linked to $G_i/G_o$ in the activation of the ras pathway[123-125] that is more commonly associated with signaling via receptor tyrosine kinases. The mechanism is obscure: it is apparently not mediated by alterations in cAMP but might be secondary to the activation of a protein kinase.[123,125]

A more direct effect of $G_i$, or $\alpha_{i3}$, has been demonstrated in mast cell secretion.[126] Neutralizing antibodies to the G protein blocked the well-documented stimulatory effect of GTP analogs on secretion from permeabilized cells. This effect had been previously postulated to be mediated by $G_e$, an undefined species that acted very late in the exocytotic pathway.[127] The exact step remains to be defined, as does its proximity and relation to the "distal site", described above, that mediates inhibition of secretion under certain circumstances.

Although the list of potential roles of G proteins is virtually endless perhaps a final role should be mentioned: stimulation of PI-3-kinase. This is the enzyme that phosphorylates $PIP_2$ to produce phosphatidylinositol 3,4,5-trisphosphate ($PIP_3$), a membrane lipid whose function is obscure, but which might involve regulation of the cytoskeleton.[128-129] The activation of this enzyme is classically thought to proceed as a consequence of tyrosine kinase cascades. However a direct, and quantitatively important contribution due to a pertussis toxin-sensitive G protein has been recently described in human neutrophils.[130]

## 6. ION CHANNELS AS EFFECTORS

In addition to interaction with enzyme effector systems, there is good evidence that G protein-coupled receptors also directly modulate various channels.[55-57] This has been mentioned above as one of the additional features of many of the receptors that couple to the inhibition of adenylyl cyclase. Regulation of channel conductance by this mechanism is slower than that associated with ligand-gated channels, but faster than that mediated by protein phosphorylation.

The first channel to be shown to be modulated in this manner was the atrial $K^+$ channel of the heart. The conductance of this channel had been known since the late 1950s to be activated by muscarinic cholinergic agonists and was shown much later to involve the m2 muscarinic receptor.[57] Evidence for direct involvement of a G protein included effects of GTP analogs which were highly localized to their point of addition, and which could not be explained by the generation of other known second messengers.[131-133] The G protein was pertussis toxin-sensitive and subsequently identified as $\alpha_{i3}$ (previously referred to as $G_k$). Purified preparations of the latter, preactivated with GTP, reproduced the effects of acetylcholine on channel activity when added directly to the membranes from which current recordings were being taken.[134] For some years it had been thought that $\beta\gamma$ subunits might also play a role in directly modulating $K^+$ channel activity,[135] but it is now generally agreed that the predominant effect is through the $\alpha$ subunit.[55,57] However no reconstitution experiments have yet been performed to verify that the interaction with $G_{i3}$ is indeed direct, and not mediated by a membrane delimited (i.e. noncytoplasmic) second messenger.[57] An example of the latter might be arachidonic acid, the generation of which by a $\beta\gamma$-stimulated activation of $PLA_2$ has been proposed to explain some of the effects of acetylcholine on the $K^+$ channel.[136] Availability of the recombinant protein, following the very recent cloning of the channel,[137] should allow resolution of some of these remaining controversies, including the possibility that the $\alpha$ and $\beta\gamma$ subunits are actually acting on different channels.[12]

In neuronal and endocrine cells pertussis toxin-sensitive G proteins are also involved in activating various $K^+$ channel conductances and inhibiting voltage-dependent $Ca^{2+}$ (and $Na^+$) channels.[53-57] Whether the coupling is direct or via a membrane delimited mediator is currently unknown. Reconstitution experiments have been performed with G protein $\alpha$ subunits and the L-type voltage-dependent $Ca^{2+}$ channels purified from heart.[138] These studies provided evidence for a direct interaction between the $\alpha_s$ and the channel. This mechanism apparently acts in addition to, but may even be implicated in,[139] the channel phosphorylation by PKA that had previously been shown to augment $Ca^{2+}$ influx. This is an interesting example of a single G protein subtype influencing a single physiological response but through both direct and indirect mechanisms.

## $\beta\gamma$ SUBUNITS

Four different $\beta$ subunits, each with a molecular weight around 37 kDa, and seven different $\gamma$ subunits (7.3-8.5 kDa) have been described in mammalian tissues.[4,24,140-141] The $\beta$ subunits are all widely expressed, as are a number of $\gamma$ subunits.[4,24,140-141] The exceptions are $\gamma_1$, chiefly expressed in photoreceptor cells, and $\gamma_2$ and $\gamma_3$, which are relatively restricted to the brain. Because of this, and the fact that some $\beta$ and $\gamma$ subunits appear incapable of association with each other,[12] the total number of $\beta\gamma$ complexes occurring in vivo is considerably less than the theoretical maximum of 28. These complexes have never been demonstrated to dissociate under physiological conditions. Specificity of the $\beta\gamma$ complex is probably governed by the $\gamma$ moiety since complexes containing identical $\beta$ subunits, but divergent $\gamma$ subunits, are associated with different functions.[12] The least controversial of these functions are firstly the maintenance of the $\alpha$ subunits in their GDP-bound, inactive state and secondly, localization of the heterotrimer to the plasma membrane. Prenylation of the $\gamma$ subunits is crucial for these roles,[142] although the exact physiological relevance of the targeting of farnesyl, as opposed to geranylgeranyl groups, to different $\gamma$ subunits is largely unknown.

Until recently evidence for a role of $\beta\gamma$ subunits as effector molecules was confined to the muscarinic $K^+$ channel where it was controversial;[135] to the retinal $PLA_2$ enzyme where it remained largely unpursued;[117] and to a yeast mating response that had no known mammalian counterpart.[143] More widespread interest has focused on the βγ subunits only as a result of their demonstrated involvement with the two big players in the G protein-coupled receptor field: adenylyl cyclase and PLC. Interaction with the various adenylyl cyclase isoforms has been briefly mentioned above. Purified βγ subunits were shown to activate synergistically the type 2 and 4 isoforms, but only in the presence of a priming dose of $G_s$-GTP. In contrast the βγ complex inhibited $G_s$-stimulated type 1 adenylyl cyclase activity.[41,144] Confirmatory results were obtained in co-transfection studies.[145]

A similar but not completely analogous picture has subsequently emerged in the PLC field.[146] The important difference is that whereas βγ subunits do stimulate certain PLC isoforms, notably PLCβ2 and PLCβ3, they appear to do so independently of any requirement for an α subunit.[75,78,80] This is despite the fact that α subunits of the $G_{16}$ class alone can activate these isoforms, albeit weakly.[76] On the other hand PLCβ1, previously identified as the target of $α_q$, is only slightly activated by βγ subunits.[79-80] As with the adenylyl cyclase molecules, the βγ subunits and the α subunits appear to interact with different regions of the PLC molecule.[147]

Although investigations of the specificity of different βγ subunits are only in their infancy it appears that they are much more promiscuous in their coupling than α subunits. Thus a number of different combinations showed similar selectivity in activating PLCβ2.[147] The principal exception is the $β_1γ_1$ complex derived from retinal cells which is less potent than other complexes in performing a number of functions in vitro.[141] More is known with regard to the inhibition of voltage-dependent $Ca^{2+}$ channels in pituitary cells. In a series of elegant experiments employing nuclear injection of antisense constructs to inhibit selective expression of a variety of α, β, and γ subunits, it was shown that coupling of the channel to an m4 muscarinic receptor was mediated by a $α_{o1}β_3γ_4$ complex, whereas coupling to a somatostatin receptor involved

$α_{o2}β_1γ_3$.[148-150] However, as only one species of $Ca^{2+}$ channel appeared to be involved, this impressive specificity seems to be directed in this instance more toward the receptor molecule, than to the effector system. Perhaps this is another function of the βγ complex, given that the α subunits are generally much more selective for different effector systems than they are for receptor species.

## RECEPTOR DESENSITIZATION

In addition to the rapid responses mediated by G protein-coupled receptors a number of more slowly developing, counter-regulatory mechanisms are also initiated. Maintenance of agonist over the period of hours to days routinely results in a decrease in the mass of one or more components of the receptor-G protein-effector complex through which the agonist is working.[151-153] These effects are mediated at the transcriptional level. Shorter to medium term agonist stimulation results in sequestration of bound receptors from the plasma membrane via endocytotic vesicles.[153] Much less is known of how this process applies to G protein-coupled receptors than for some other receptor types such as the insulin receptor. The most rapid of the counter-regulatory mechanisms is known as receptor desensitization and develops over a time of course of seconds.[153-154] This involves at least two processes, heterologous and homologous desensitization, both of which involve phosphorylation of key components of the coupling pathway.[153]

Heterologous desensitization was originally described for the adenylyl cyclase system but applies equally well to PLC coupling. The usual consequence of stimulation of adenylyl cyclase is the activation of PKA. As it happens receptors coupling to $G_s$ are invariably good substrates for PKA, and phosphorylation of the receptor inhibits its ability to couple.[153,155] This is known as heterologous desensitization because the PKA does not discriminate between the species of receptor which originally generated the activation signal and other species of $G_s$-coupled receptor which may or may not be occupied by their cognate ligands. Glucagon for example brings about desensitization of not only glucagon receptors but β-adrenergic ones as well. An analogous role is played by PKC in the PLC system, except that in this case the relevance of phosphorylation of the receptor, as opposed to other potential targets, is less clear. In any event the functional coupling of the receptor to PLC is diminished. The heterologous nature of this phenomenon is suggested by the widespread observations that exogenous PKC activators inhibit coupling through virtually all $G_q$-coupled receptors and that agonist-stimulated inositol phosphate generation is often larger and more prolonged in PKC-depleted cells.[156-158] Heterologous desensitization, at least in its PKA-mediated guise, requires only low agonist concentrations for its initiation.[153,155]

The understanding of homologous desensitization has arisen largely as a result of the study of light adaptation in photoreceptor cells. As described above, signaling through the rhodopsin phosphodiesterase system results eventually in the inhibition of the cation-specific channel. It is vital that this channel be rapidly reopened in response to a decrease in illumination. In contrast to virtually all other G protein-coupled receptors, where a decrease in agonist concentration is rapidly detected, deactivation of photoreceptor is not limited by the removal of the light stimulus, but by the reconversion of the active chromophore all-*trans*-retinal back to the inactive 11-*cis* form, which takes several minutes and involves initial dissociation of the chromophore from the apoprotein, opsin.[60] Post-receptor mechanisms are therefore crucial for counter-regulation of the response. This is achieved by phosphorylation of rhodopsin by a specific rhodopsin kinase enzyme.[159] Most importantly only the active form of rhodopsin serves as a substrate; most probably the phosphorylation sites are occluded by transducin in the resting state. Rhodopsin kinase is a cytosolic enzyme which translocates to the plasma membrane upon stimulation with

light. The underlying mechanism is not completely understood, but is thought to involve isoprenylation of the C-terminus of the enzyme.[160] Phosphorylation of rhodopsin does not directly inhibit its recoupling with transducin. Rather it leads to the recruitment of a third protein, known as arrestin, which is thought to act physically to prevent reassociation of rhodopsin and transducin.[161]

The discovery in the late 1980s of homologs of rhodopsin kinase displaying much broader tissue distributions, led to an appreciation that similar processes might operate in cells other than photoreceptors.[162-164] Since the substrates of these rhodopsin kinase homologs were the β-adrenergic receptors they were originally christened β-adrenergic receptor kinases (βARKs). Like rhodopsin kinase these only phosphorylated the activated forms of their substrate receptors, which in turn promoted association with a third protein, the arrestin homolog, β-arrestin.[163-165] PKA and βARK phosphorylated different sites on the receptor, appeared to act independently, and made equivalent contributions to receptor desensitization at high agonist concentrations, although PKA was more active during lower levels of receptor occupation.[153] The substrate specificity of the βARK enzymes appears to be less restrictive than for rhodopsin kinase, since βARK1 has recently been shown to be identical to a kinase previously identified[166] on the basis of its ability to phosphorylate muscarinic receptors.[167] The βARK enzymes are not prenylated and intriguingly their translocation to the plasma membrane appears to require association with G protein βγ subunits which, of course, do contain this lipid modification.[166,168] This was an exciting finding because it implied that a receptor kinase of broad specificity might be precisely targeted to an activated receptor by virtue of the functional dissociation of the G protein complex. Thus a limited number of kinases might underlie the heterologous desensitization of not only all known $G_s$-coupled receptors, but possibly $G_q$-coupled ones also. However the situation is clearly not that simple since additional members of the G protein receptor kinase family have been very recently cloned that are neither prenylated themselves nor associate with $\beta\gamma$ subunits.[169-170] A better understanding of the function and mechanism of action of these enzymes may shed light on the process of heterologous desensitization which occurs in receptors linked to PLC. Little is known of this except that these receptors are phosphorylated as a result of ligand binding and in a manner not wholly attributable to PKC.[154]

# SPECIFICITY, DIVERSIFICATION AND INTEGRATION OF SIGNALING PATHWAYS

Occupation of the hundreds of different species of G protein-coupled receptor leads to hundreds of different physiological responses throughout the body. Yet the number of G proteins is limited and the number of effector systems even more so. How do cells integrate the information from diverse ligands without losing the specificity of the original signal, and how do they differentiate the multiple signaling read-outs which are derived from a handful of effector systems? The answers to these questions are poorly understood and likely to be complex. Nevertheless some preliminary explanations can now be proffered.

One contributing factor would be the presence on a single cell of different receptor subtypes capable of binding a single agonist species. Graded responses would occur depending on the ratio of the various receptor subtypes. Although undoubtedly a contributing factor, interactions between receptor subtypes could be considered merely as part of the wider question of interaction between multiple receptor classes. In any event, post receptor-mechanisms have been clearly implicated in the generation of signaling diversity from the results of transfection studies in which cDNAs encoding a single receptor subspecies were used. These demonstrated that the heterologously expressed receptor could activate more than one signaling pathway

in a single cell,[171-172] and sometimes completely different pathways depending on the exact cell type employed as a transfection recipient.[173] The key questions are therefore, can a single receptor interact with more than one G protein; and can a single G protein activate more than one effector system?

The transfection experiments described above only begin to address this issue because they set the upper limit for the number of such interactions, and may or may not give an indication of which interactions would be physiologically relevant. The specificity of interactions in the receptor-G protein-effector complex have therefore been addressed in various additional ways. One of the most powerful techniques involved the purification of a particular receptor from a biological source and identification by immunoblotting of the G proteins co-purifying with it.[73,174] This pointed, for example, to the role of $G_q$ in the activation of PLC, a role subsequently confirmed by a second technique: reconstitution of various purified components into a functioning signaling complex.[72,175-176] A third method is the use of neutralizing antibodies to disrupt specifically the interaction of a single class of G protein $\alpha$ subunit with either its receptor,[177-178] or effector enzyme,[179-180] in cell-free systems. Photoaffinity labeling using GTP analogs has also been employed to identify which $\alpha$ subunit is activated in response to a particular agonist.[181-182] The fifth technique, injection of antisense constructs to attenuate functional expression of selected G protein subunits has already been described.[148] While each of these techniques has its own strengths and weaknesses and, indeed, has sometimes given conflicting results when applied to the same system, collectively they have led to the various assignations of particular G proteins to the defined functions described above. They also provide very strong evidence that multiple coupling between single receptors and various G proteins is physiologically relevant. This is especially evident for those receptors coupling to inhibition of adenylyl cyclase through $\alpha_{i2}$.[174,179,181-182] As described above these receptors invariably interact with other $\alpha_o/\alpha_i$ subunits and thereby trigger diverse signaling pathways. Whether signal diversification can be initiated by a single G protein $\alpha$ subunit in a defined cell type is still somewhat controversial, although a given $\alpha$ species might exert different roles in different cell types. Perhaps the best evidence for multiple interaction comes from cardiac cells where $\alpha_q$ appears to interact with $Ca^{2+}$ channels as well as adenylyl cyclase. However, since the channel is also a substrate for PKA, $G_s$ in this case achieves signal reinforcement, rather than diversification, by using the two effector systems. The demonstration that $\beta\gamma$ subunits also exert signaling function is of great significance in any attempt to explain the generation of multiple signals from a single G protein complex. Thus in hemopoietic cells, the effect of a single receptor species both to inhibit adenylyl cyclase and to activate PLC in a pertussis toxin-sensitive fashion, is most probably due to the respective actions of the $\alpha$ and $\beta\gamma$ subunits derived from a single $G_i$ species.

## CROSS-TALK

The previous discussion has implied that signaling proceeds in a fashion predetermined solely by the structure of the receptor to which a ligand has bound. In reality the physiological response to a given agonist will be determined as a function of a host of other influences. For example, the expression of various effector molecules might be developmentally regulated or controlled by the nutritional status of the cell. Alternatively, the response might be downregulated as a result of prior exposure to the ligand. Finally, an agonist does not act in isolation, even over the short term, and hence the final physiological event will represent the integration of all the signaling pathways operating at any one time. One example, already discussed, is how small inputs from diverse receptor species might be summed at the level of a single G protein subunit to bring about

the significant activation of an effector system. Alternatively the cross-talking between multiple receptor species might be considerably more complex than simple summation. Examples of these more complex interactions have also been discussed previously and would include heterologous desensitization, and modulation of stimulated adenylyl cyclase activity by $\beta\gamma$ subunits derived from G proteins of the $G_i/G_o$ family. Other functionally documented cross-talk responses include those where a receptor generally linked to inhibition of adenylyl cyclase can, in the presence of an activated $G_q$, potentiate PLC.[183-184] In these instances $\beta\gamma$ subunits are also very probably involved, although there is as yet no direct evidence.

Phosphorylation is another major mechanism for mediating cross-talk between G protein-coupled receptors, as has been outlined above for receptor desensitization. However other targets acting more distally to the receptor in the signaling sequences have also been proposed. The most obvious are G protein subunits, as in the phosphorylation of $\alpha_i$[185] and $\alpha_z$[186] by PKC, although the physiological relevance of these events has not yet been conclusively proved. Perhaps surprisingly there is very little evidence for effector systems acting as substrates for protein kinases, except for $Ca^{2+}$ channels which are usually activated by PKA and inactivated by PKC. An interesting exception occurs in pituitary cells in which $G_{i2}$ activates $Ca^{2+}$ influx only if the channel is phosphorylated by PKC.[187] The $IP_3$ receptor is also phosphorylated by PKA, resulting in either decreased or increased ability to release $Ca^{2+}$ in brain[86,188] and liver, respectively.[189] Enzymes involved in the removal of second messengers are also highly regulated and potential sites of interaction with other signaling pathways. The cyclic AMP/GMP phosphodiesterases are variously modulated by PKC, PKA and calmodulin, while the latter is well-documented to augment the activity of $IP_3$ kinase.[190] Finally a large number of kinase substrates are subjected to phosphorylation on multiple sites depending on the exact kinase involved, and their order of action. Such hierarchical phosphorylation is thought to play an important role in a number of specialized physiological processes such as secretion and muscle contraction but is at present only poorly understood.[191]

## CONCLUSIONS

Until very recently the greatest puzzle to those interested in the signaling pathways deriving from G protein-coupled receptors was how the diverse array of physiological responses, triggered by receptor binding, could be mediated by such a limited number of G proteins. Although many details remain to be filled in, this conceptual difficulty has largely been resolved, firstly by the molecular cloning experiments that showed a greater diversity of G protein species than had been previously envisioned, and secondly by the growing evidence that $\beta\gamma$ subunits are capable of interaction with effector systems. The most obvious next step is to catalog which subunit species are involved in the coupling to which isoform of receptor and effector system. An equally obvious goal, but one which may prove very difficult to attain, is the identification of the functions of $G_z$ and $G_{12/13}$.

However, perhaps the chief conceptual problem which needs to be resolved is how signals from different G protein-coupled receptors are spatially integrated and disseminated. For example, potentiation of Type 2 adenylyl cyclase activity can be achieved through interaction of the $\alpha$ subunit of $G_s$ and $\beta\gamma$ derived from $G_i$. Since $\alpha$ and $\beta\gamma$ subunits are unlikely to be freely diffusible, even if they were capable of complete dissociation, such a scenario presupposes close apposition of $G_s$, $G_i$ and Type 2 adenylyl cyclase. Moreover, under different circumstances $\beta\gamma$ subunits might either inhibit adenylyl cyclase or activate PLC, or a given $\alpha_i$ species might interact with either adenylyl cyclase or an ion channel. Recruitment of effector molecules, and the clustering of G proteins into functional arrays, as are implied by these results, can only be explained by some form of spatial regulation, almost certainly involving the cytoskeleton. Some evidence already exists for association of receptors, G proteins and effectors with the cytoskeleton in human neutrophils,[192] and is probably involved in the restriction of $\alpha_{i3}$ to the apical, as opposed to basolateral, membranes of renal epithelial cells.[193] In lymphoma cells activation of adenylyl cyclase by $G_s$ is inhibited by disruptors of microtubules, whereas coupling through $G_i$ is unaffected.[194] Further work along these lines will not only shed light on receptor-effector coupling, but might also help determine whether G proteins have been seconded to perform other roles in the cell, at sites far removed from the plasma membrane and G protein-coupled receptors. Such roles might include the regulation of neuronal growth by $G_o$, which in this instance appears to be controlled not by a membrane receptor, but by the cytosolic protein GAP-43.[195] It is therefore probable that G protein-coupled receptors, as widespread and important as they are, might not be the sole means of activating G proteins.

## REFERENCES

1. Bourne HR, Sanders DA, McCormick F. The GTPase superfamily: conserved structure and molecular mechanism. Nature 1991; 349:117-127.
2. Obar RA, Collins CA, Hammarback JA et al. Molecular cloning of the microtubule-associated mechanochemical enzyme dynamin reveals homology with a new family of GTP-binding proteins. Nature 1990; 347:256-261.
3. Gilman AG. G proteins and regulation of adenylyl cyclase. J Am Med Assoc 1989; 262:1819-1825.
4. Hepler JR, Gilman AG. G proteins. Trends Biochem Sci 1992; 17:383-387.
5. Spiegel AM. G proteins in cellular control. Curr Opin Cell Biol 1992; 4:203-211.
6. Cassel D, Selinger Z. Catecholamine-stimulated GTPase activity in turkey erythrocytes. Biochim Biophys Acta 1976; 452:538-551.
7. Maguire ME, Van Arsdale PMV, Gilman AG. An agonist-specific effect of guanine nucleotides on binding to the β-adrenergic

receptor. Mol Pharmacol 1976; 12:335-339.

8. Schramm M, Rodbell M. A persistent active state of the adenylate cyclase system produced by the combined actions of isoproterenol and guanylyl imidodiphosphate in frog erythrocyte membranes. J Biol Chem 1975; 250:2232-2237.

9. Cassel D, Selinger Z. Mechanism of adenylate cyclase activation by cholera toxin: an inhibition of GTP hydrolysis at the regulatory site. Proc Natl Acad Sci USA 1977; 74:3307-3311.

10. Katada T, Ui M. Direct modification of the membrane adenylate cyclase system by islet-activating protein due to ADP-ribosylation of a membrane protein. Proc Natl Acad Sci USA 1982; 79:3129-3133.

11. Rodbell M. G-protein alpha subunits as programmable second messengers. Trends Biochem Sci 1985; 10:461-465.

12. Clapham DE, Neer EJ. New roles for G-protein $\beta\gamma$-dimers in transmembrane signalling. Nature 1993; 365:403-406.

13. Yi F, Denker BM, Neer EJ. Structural and functional studies of cross-linked $G_o$ protein subunits. J Biol Chem 1991; 266:3900-3906.

14. Stow JL, Almeida JBD, Narula N et al. A heterotrimeric G protein, $G\alpha_{i-3}$, on golgi membranes regulates the secretion of a heparan sulfate proteoglycan in LLC-PK1 epithelial cells. J Cell Biol 1991; 114:1113-1124.

15. Donaldson JG, Kahn RA, Lippincott-Schwartz J et al. Binding of ARF and beta-COP to golgi membranes: possible regulation by a trimeric G protein. Science 1991; 254:1197-1199.

16. Crouch MF. Growth factor-induced cell division is paralleled by translocation of $G_i$ alpha to the nucleus. FASEB J 1991; 5:200-206.

17. Taylor CW. The role of G proteins in transmembrane signalling. Biochem J 1990; 272:1-13.

18. Gilman AG. G proteins: transducers of receptor-generated signals. Annu Rev Biochem 1987; 56:615-649.

19. Birnbaumer L. Receptor-to-effector signalling through G proteins: roles for $\beta\gamma$ dimers as well as $\alpha$ subunits. Cell 1992;

71:1069-1072.

20. Milligan C. Mechanisms of multifunctional signalling by G protein linked receptors. Trends Pharmacol Sci 1993; 14:239-244.

21. Berstein G, Blank JL, Jhon DY et al. Phospholipase C beta-1 is a GTPase-activating protein for Gq/11 its physiological regulator. Cell 1992; 70:411-418.

22. Arshavsky, Bownds MD. Regulation of deactivation of photoreceptor G protein by its target enzyme and cGMP. Nature 1992; 357:416-417.

23. Bauer PH, Muller S, Puzicha M et al. Phosducin is a protein kinase A-regulated G-protein regulator. Nature 1992; 358:73-76.

24. Simon MI, Strathmann MP, Gautam N. Diversity of G proteins in signal transduction. Science 1991; 252:802-808.

25. Casey PJ, Fong HKW, Simon MI et al. $G_z$, a guanine nucleotide-binding protein with unique biochemical properties. J Biol Chem 1990; 265:2383-2390.

26. Landis CA, Masters SB, Spada A. GTPase inhibiting mutations activate the $\alpha$ chain of $G_s$ and stimulate adenylyl cyclase in human pituitary tumours. Nature 1989; 340:692-696

27. Lyons J, Landis CA, Harsh G et al. Two G protein oncogenes in human endocrine tumours. Science 1990; 249:655-659.

28. Weinstein LS, Shenker A, Gejman PV et al. Activating mutations of stimulatory G protein in the McCune-Albright Syndrome. New Engl J Med 1991; 325:1688-1695.

29. Journot L, Pantaloni C, Bockaert J et al. Deletion within the amino terminal region of $G_s\alpha$ impairs its ability to interact with $\beta\gamma$ subunits and to activate adenylate cyclase. J Biol Chem 1991; 266:9009-9015.

30. Jones TLZ, Simonds WF, Merendino JJ et al. Myristoylation of an inhibitory GTP binding protein $\alpha$ subunit is essential for its membrane attachment. Proc Natl Acad Sci USA 1990; 87:568-572.

31. Linder ME, Pang I-H, Duronio RJ et al. Lipid modifications of G protein subunits. Myristoylation of $G_o\alpha$ increases its affinity for $\beta\gamma$. J Biol Chem 1991; 266:4654-4659.

32. Kokame K, Fukada Y, Yoshizawa T. Lipid modification at the N terminus of photore-

ceptor G-protein α-subunit. Nature 1992; 259:671-672.

33. Audiger Y, Journot L, Pantaloni C et al. The carboxy terminal domain of G$_s$α is necessary for anchorage of the activated form to the plasma membrane. J Cell Biol 1990; 111:427-1435.

34. Noel JP, Hamm HE, Sigler PB. The 2.2Å crystal structure of transduction-α complexed with GTPγS. Nature 1993; 366: 654-663.

35. Markby DW, Onrust R, Bourne HR. Separate GTP binding and GTPase activating domains of a Gα subunit. Science 1993; 262:1895-1901.

36. Rall TW, Sutherland EW. The regulatory role of adenosine-3',5'-phosphate. Cold Spring Harbor Symp Quant Biol 1961; 26:347-354.

37. Limbird LE, Lefkowitz RJ. Resolution of β-adrenergic receptor binding and adenylate cyclase activity by gel exclusion chromotography. J Biol Chem 1977; 252:799-802.

38. Rodbell M, Birnbaumer L, Pohl SL et al. The glucagon-sensitive adenyl cyclase system in plasma membranes of rat liver. V. An obligatory role of guanylnucleotides in glucagon action. J Biol Chem 1971; 246:1877-1882.

39. Ross EM, Howlett AC, Ferguson KM et al. Reconstitution of hormone sensitive adenylate cyclase activity with resolved components of the enzyme. J Biol Chem 1978; 253:6401-6412.

40. Sternweis PC, Northup JK, Smigel MD et al. The regulatory component of adenylate cyclase. Purification and properties. J Biol Chem 1981; 256:11517-11526.

41. Tang W-J, Gilman AG. Adenylyl cyclases. Cell 1992; 70:869-872.

42. Pieroni JP, Jacobowitz O, Chen J et al. Signal recognition and integration by G$_s$-stimulated adenylyl cyclases. Curr Opin Neurobiol 1993; 3:345-351.

43. Krebs EG. Role of cyclic AMP-dependent protein kinase in signal transduction. J Am Med Assoc 1989; 262:1815-1818.

44. Adams SR, Harootunian AI, Buechler YJ et al. Flourescence ratio imaging of cyclic AMP in single cells. Nature 1991;

349:694-697.

45. Cohen P. The structure and regulation of protein phosphatases. Annu Rev Biochem 1989; 58:453-508.

46. McKnight GS, Cadd GG, Clegg CH et al. Expression of wild-type and mutant subunits of the cAMP-dependent protein kinase. Cold Spring Harbor Symp Quant Biol 1988; 53:111-119.

47. McKnight GS. Cyclic AMP second messenger systems. Curr Opin Cell Biol 1991; 3:213-217.

48. Scott JD, Soderling TR. Serine/theonine protein kinases. Curr Opin Neurobiol 1992; 2:289-295.

49. Beebe SJ, Oyen O, Sandberg M et al. Molecular cloning of a tissue-specific protein kinase (Cγ) from human testis - representing a third isoform for the catalytic subunit of cAMP-dependent protein kinase. Mol Endocrinol 1990; 4:465-475.

50. Itoh H, Kozasa T, Nagata S. Molecular cloning and sequence determination of cDNAs for α subunits of the guanine nucleotide-binding proteins G$_s$, G$_i$ and G$_o$ from rat brain. Proc Natl Acad Sci USA 1986; 83:3776-3780.

51. Chen J, Iyengar R. Inhibition of cloned adenylyl cyclases by mutant activated G$_i$α and specific suppression of type 2 adenylyl cyclase inhibition by phorbol ester treatment. J Biol Chem 1993; 268:12253-12256.

52. Limbird LE. Receptors linked to inhibition of adenylate cyclase: additional signalling mechanisms. FASEB J 1988; 2:2686-2695.

53. Garcia-Sainz JA. "Inhibitory" receptors and ion channel effectors. Trends Pharmacol Sci 1988; 9:27-28.

54. Vallar L, Meldolesi J. Mechanism of signal transduction at the dopamine D$_2$ receptor. Trends Pharmacol Sci 1989; 10:74-77.

55. Hille B. G protein-coupled mechanisms and nervous signalling. Neuron 1992; 9:187-195.

56. Hescheler J, Schultz G. G-proteins involved in the calcium channel signalling systems. Curr Opin Neurobiol 1993; 3:360-367.

57. Brown AM. Membrane-delimited cell signalling complexes: direct ion channel regulation by G-proteins. J Membr Biol 1993;

131:93-104.

58. Ullrich S, Wollheim CB. GTP-dependent inhibition of insulin secretion by epinephrine in permeabilized RINm5F cells: lack of correlation between insulin secretion and cyclic AMP levels. J Biol Chem 1988; 263:8615-8620.

59. McDermott AM, Sharp GWG. Inhibition of insulin secretion: a fail-safe system. Cell Signal 1993; 5:229-234.

60. Stryer L. Visual excitation and recovery. J Biol Chem 1991; 266:10711-10714.

61. Bentley JK, Beavo JA. Regulation of cyclic nucleotides. Curr Opin Cell Biol 1992; 4:233-240.

62. Berridge MJ, Irvine RF. Inositol trisphosphate, a novel second messenger in cellular signal transduction. Nature 1984; 312:315-321.

63. Berridge MJ, Irvine RF. Inositol phosphates and cell signalling. Nature 1989; 341:197-205.

64. Lassing I, Lindberg U. Specific interaction between phosphatidylinositol 4,5-bisphosphate and profilactin. Nature 1985; 314:472-474.

65. Goldschmidt-Clermont PJ, Machesky LM, Baldassare JJ et al. The actin-binding protein prolifin binds to $PIP_2$ and inhibits its hydrolysis by phospholipase C. Science 1990; 247:1575-1578.

66. Rhee SG, Choi K. Regulation of inositol phospholipid-specific phospholipase C isozymes. J Biol Chem 1992; 267: 12393-12396.

67. Cockcroft S, Thomas GMH. Inositol-lipid-specific phospholipase C isoenzymes and their differential regulation by receptors. Biochem J 1992; 288:1-14.

68. Dixon JF, Hokin LE. Kinetic analysis of the formation of inositol 1:2-cyclic phosphate in carbachol-stimulated pancreatic minilobules: half is formed by direct phosphodiesteratic cleavage of phosphatidylinositol. J Biol Chem 1989; 264:11721-11724.

69. Biden TJ, Prugue ML, Davidson AGM. Evidence for phosphatidylinositol hydrolysis in islets stimulated with carbamoylcholine. Kinetic analysis of inositol polyphosphate metabolism. Biochem J 1992;

285:541-549.

70. Haslam RJ, Davidson ML. Receptor-induced diacylglycerol formation in permeabilized platelets; possible role for a GTP-binding protein. J Recept Res 1984; 4:1-6.

71. Fain JN, Wallace MA, Wojcikiewicz RJH. Evidence for involvement of guanine nucleotide binding regulatory proteins in the activation of phospholipases by hormones. FASEB J 1988; 2:2569-2574.

72. Smrcka AV, Helper JR, Brown KO et al. Regulation of polyphosphoinositide-specific phospholipase C activity by purified $G_q$. Science 1991; 251:804-807.

73. Taylor SJ, Chae HZ, Rhee SG et al. Activation of the β1 isozyme of phospholipase C by α subunits of the $G_q$ class of G proteins. Nature 1991; 350:516-518.

74. Strathmann M, Simon MI. G protein diversity: a distinct class of α subunits is present in vertebrates and invertebrates. Proc Natl Acad Sci USA 1990; 87:9113-9117.

75. Park D, Jhon D-Y, Kriz R et al. Cloning, sequencing, expression, and $G_q$-independent activation of phospholipase C-β2. J Biol Chem 1992; 267:16048-16055.

76. Jhon D-Y, Lee H-H, Park D et al. Cloning, sequencing, purification, and $G_q$-dependent activation of phospholipase C-β3. J Biol Chem 1993; 268:6654-6661.

77. Kim MJ, Bahk YY, Min DS et al. Cloning of cDNA encoding rat phospholipase C-β4, a new member of the phospholipase C. Biochem Biophys Res Commun 1993; 194:706-712.

78. Camps M, Carozzi A, Schnabel P et al. Isozyme-selective stimulation of phospholipase C-β2 by G protein βγ-subunits. Nature 1992; 360:684-686.

79. Park D, Jhon D-Y, Lee C-W et al. Activation of phospholipase C isozymes by G protein βγ subunits. J Biol Chem 1993; 268:4573-4576.

80. Smrcka AV, Sternweis PC. Regulation of purified subtypes of phosphatidylinositol-specific phospholipase C β by G protein and βγ subunits. J Biol Chem 1993; 268:9667-9674.

81. Berridge MJ. Inositol trisphosphate and calcium signalling. Nature 1993; 361: 315-325.

82. Spat A, Bradford PG, McKinley JS et al. A saturable receptor for $^{32}$P-inositol-1,4,5-trisphosphate in hepatocytes and neutrophils. Nature 1986; 319:514-516.

83. Furuichi T, Yoshikawa S, Miyawaki A et al. Primary structure and functional expression of the inositol 1,4,5-trisphosphate-binding protein $P_{400}$. Nature 1989; 342:32-38.

84. Sudhof TC, Newton CL, Archer BT et al. Structure of a novel InsP$_3$ receptor. EMBO J 1991; 10:3199-3206.

85. Blondel O, Takeda J, Janssen H et al. Sequence and functional characterisation of a third inositol trisphosphate receptor subtype, IP3R-3, expressed in pancreatic islets, kidney, gastrointestinal tract, and other tissues. J Biol Chem 1993; 268:11356-11363.

86. Danoff SK, Ferris CD, Donath C et al. Inositol 1,4,5-trisphosphate receptors: Distinct neuronal and nonneuronal forms derived by alternative splicing differ in phosphorylation. Proc Natl Acad Sci USA 1991; 88:2951-2955.

87. Streb HP, Irvine RF, Berridge MJ et al. Release of Ca$^{2+}$ from a nonmitochondrial store in pancreatic acinar cells by inositol-1,4,5-trisphosphate. Nature 1983; 306:67-69.

88. Means AR, VanBerkum MF, Bagchi I et al. Regulatory functions of calmodulin. Pharmacol Therap 1991; 50:255-270.

89. Irvine RF, Moor RM, Pollock WK et al. Inositol phosphates: proliferation, metabolism and function. Phil Trans R Soc Lond Biol 1988; 320:281-298.

90. Putney JW. A model for receptor-regulated Ca$^{2+}$ entry. Cell Calcium 1986; 7:1-12.

91. Penner R, Fasolato C, Hoth M. Calcium influx and its control by calcium release. Curr Opin Neurobiol 1993; 3:368-374.

92. Randriamampita C, Tsien RY. Emptying of intracellular Ca$^{2+}$ stores releases a novel small messenger that stimulates Ca$^{2+}$ influx. Nature 1993; 364:809-814.

93. Fasolato C, Hoth M, Penner R. A GTP-dependent step in the activation mechanism of capacitative calcium influx. J Biol Chem 1993; 268:20737-20740.

94. Bird GS, Putney JW. Inhibition of thapsigargin-induced calcium entry by microinjected guanine nucleotide analogues. Evidence for the involvement of a small G-protein in capacitative calcium entry. J Biol Chem 1993; 268:21486-21488.

95. Shears SB. Metabolism of the inositol phosphates produced upon receptor activation. Biochem J 1989; 260:313-324.

96. Majerus PW. Inositol phosphate biochemistry. Annu Rev Biochem 1992; 61:225-250.

97. Menniti FS, Oliver KG, Putney JW et al. Inositol phosphates and cell signalling: new views of Ins P$_5$ and Ins P$_6$. Trends Biochem Sci 1993; 18:53-56.

98. Takai Y, Kishimoto A, Inoue M et al. Studies on a cyclic nucleotide-independent protein kinase and its proenzyme in mammalian tissues. J Biol Chem 1977; 252:7603-7609.

99. Nishizuka Y. Intracellular signalling by hydrolysis of phospholipids and activation of protein kinase C. Science 1992; 258:607-614.

100. Osada S, Mizuno K, Saido TC et al. A new member of the protein kinase C family, nPKCθ, specifically expressed in skeletal muscle. Mol Cell Biol 1992; 12:3930-3938.

101. Selbie LA, Schmitz-Peiffer C, Sheng Y et al. Molecular cloning and characterisation of PKCι, an atypical isoform of protein kinase C derived from insulin-secreting cells. J Biol Chem 1993; 268:24296-24302.

102. Castana M, Takai Y, Kaibuchi K et al. Direct activation of calcium-activated, phospholipid-dependent protein kinase by tumor-promoting phorbol esters. J Biol Chem 1982; 257:7847-7851.

103. Kraft AS, Anderson WB. Phorbol esters increase the amount of Ca$^{2+}$, phospholipid-dependent protein kinase associated with plasma membrane. Nature 1983; 301:621-623.

104. Wolf M, Cautrecasas P, Sahyoun N. Interaction of protein kinase C with membranes is regulated by Ca$^{2+}$, phorbol esters, and ATP. J Biol Chem 1985; 260:15718-15722.

105. Parker PJ, Coussens L, Totty N et al. The complete primary structure protein kinase C—the major phorbol ester receptor. Science 1986; 233:853-859.

106. Coussens L, Parker PJ, Rhee L et al. Multiple distinct forms of bovine and human

protein kinase C suggest diversity in cellular signalling pathways. Science 1986; 233:859-866.

107. Bell RM, Burns DJ. Lipid activation of protein kinase C. J Biol Chem 1991; 266:4661-4664.

108. Mochly-Rosen D, Khaner H, Lopez J. Identification of intracellular receptor proteins for activated protein kinase C. Proc Natl Acad Sci USA 1991; 88:3997-4000.

109. Trilivas I, McDonough PM, Brown JH. Dissociation of protein kinase C redistribution from the phosphorylation of its substrates. J Biol Chem 1991; 266:8431-8438.

110. Hug H, Sarre TF. Protein kinase C isoenzymes: divergence in signal transduction? Biochem J 1993; 291:329-343.

111. Shears SB. Regulation of the metabolism of 1,2-diaylglycerol and inositol phosphates that respond to receptor activation. Pharmac Therap 1991; 49:79-104.

112. Schimmel RJ. The $\alpha_1$-adrenergic transduction system in hamster brown adipocytes: release of arachidonic acid accompanies activation of phospholipase C. Biochem J 1988; 253:93-102

113. Balsinde J, Diez E, Mollinedo F. Arachidonic acid release from diacylglycerol in human neutrophils. Translocation of diacylglycerol deacylating enzyme activities from an intracellular pool to plasma membrane upon cell activation. J Biol Chem 1991; 266:15638-15643.

114. Linden J, Delahunty TM. Receptors that inhibit phosphoinositide beakdown. Trends Pharmacol Sci 1989; 10:112-120.

115. Bizzarri C, DiGirolamo M, D'Orazio MC et al. Evidence that a guanine nucleotide-binding protein linked to a muscarinic receptor inhibits directly phospholipase C. Proc Natl Acad Sci USA 1990; 87:4889-4893.

116. Litosch I, Sulkholutskaya I, Weng C. G protein-mediated inhibition of phospholipase C activity in a solubilized membrane preparation. J Biol Chem 1993; 268:8692-8697.

117. Axelrod J. Receptor-mediated activation of phospholipase $A_2$ and arachidonic acid release in signal transduction. Biochem Soc Trans 1990; 18:503-507.

118. Clark JD, Lin L-L, Kriz RW et al. A novel arachidonic acid-selective cytosolic $PLA_2$ contains a $Ca^{2+}$-dependent translocation domain with homology to PKC and GAP. Cell 1991; 86:1043-1051.

119. Chan K-M, Turk J. Mechanism of arachidonic acid-induced $Ca^{2+}$ mobilization from rat liver microsomes. Biochem Biophys Acta 1987; 928:186-193.

120. Billah MM, Anthes JC. The regulation and cellular functions of phosphatidylcholine hydrolysis. Biochem J 1990; 269:281-291.

121. Exton JH. Signalling through phosphatidylcholine breakdown. J Biol Chem 1990; 265:1-4.

122. Pan MG, Florio T, Stork PJS. G protein activation of a hormone-stimulated phosphatase in human tumor cells. Science 1993; 256:1215-1217.

123. van Corven EJ, Hordijk PL, Medema RH et al. Pertussis toxin-sensitive activation of $p21^{ras}$ by G protein-coupled receptor agonists in fibroblasts. Proc Natl Acad Sci USA 1993; 90:1257-1261.

124. Winitz S, Russell M, Qian N-X et al. Involvement of Ras and Raf in the $G_i$-coupled acetylcholine muscarinic m2 receptor activation of mitogen-activated protein (MAP) kinase kinase and MAP kinase. J Biol Chem 1993; 268:19196-19199.

125. Alblas J, van Corven EJ, Hordijk PL et al. $G_i$-mediated activation of the $p21^{ras}$-mitogen-activated protein kinase pathway by $\alpha_2$-adrenergic receptors expressed in fibroblasts. J Biol Chem 1993; 268:22235-22238.

126. Aridor M, Rajmilevich G, Beaven MA et al. Activation of exocytosis by the heterotrimeric G protein $G_{i3}$. Science 1993; 262:1569-1572.

127. Comperts BD. $G_e$: a GTP-binding protein mediating exocytosis. Annu Rev Physiol 1990; 52:591-606.

128. Carpenter CL, Duckworth BC, Auger KR et al. Purification and characterisation of phosphatidylinositol 3-kinase from bovine thymus: monomer and heterodimer forms. J Biol Chem 1990; 265:19704-19711.

129. Stephens LR, Hughes KT, Irvine RF. Pathway of phosphatidylinositol (3,4,5)-trisphosphate synthesis in activated neutrophils. Nature 1991; 351:33-38.

130. Stephens LR, Jackson T, Hawkins PT. Syn-

thesis of phosphatidylinositol (3,4,5)-trisphosphate synthesis in permeabilized neutrophils regulated by receptors and G-proteins. J Biol Chem 1993; 268:17162-17172.

131. Pfaffinger PJ, Martin JM, Hunter DD et al. GTP-binding proteins couple cardiac muscarinic receptors to a K channel. Nature 1985; 317:536-538.

132. Breitwieser GE, Szabo G. Uncoupling of cardiac muscarinic and β-adrenergic receptors from ion channels by a guanine nucleotide analogue. Nature 1985; 317:538-540.

133. Soejima M, Noma A. Mode of regulation of the ACh-sensitive K channel by the muscarinic receptor in rabbit atrial cells. Pflugers Arch 1984; 400:424-431.

134. Yatani A, Mattera R, Codina J et al. The G protein-gated atrial K⁺ channel is stimulated by three distinct Gᵢα subunits. Nature 1988; 336:680-682.

135. Logothetis DE, Kurachi Y, Galper J et al. The βγ-subunits of GTP-binding proteins activate the muscarinic K⁺ channel in heart. Nature 1987; 325:321-326.

136. Kim D, Lewis DL, Grazidei L et al. G-protein βγ-subunits activate the cardiac muscarinic K⁺-channel via phospholipase A₂. Nature 1989; 337:557-560.

137. Kubo Y, Reuveny E, Slesinger PA et al. Primary structure and functional expression of a rat G-protein-coupled muscarinic potassium channel. Nature 1993; 364:802-806.

138. Yatani A, Codina J, Imoto Y et al. A G protein directly regulates mammalian cardiac calcium channels. Science 1987; 238:1288-1292.

139. Cavalie A, Allen TJ, Trautwein W. Role of the GTP-binding protein G₅ in the β-adrenergic modulation of cardiac Ca channels. Pflugers Arch 1991; 419:433-443.

140. Cali JJ, Balcueva EA, Rybalkin I et al. Selective tissue distribution of G protein γ subunits, including a new form of the γ subunits identified by cDNA cloning. J Biol Chem 1992; 267:24023-24027.

141. Asano T, Morishita R, Matsuda T et al. Purification of four forms of the βγ subunit complex of G proteins containing different γ subunits. J Biol Chem 1993; 268:20512-20519.

142. Simonds WF, Butrynski JE, Gautam N et al. G protein beta/gamma dimers: Membrane targeting requires subunit coexpression and intact gamma CAAX domain. J Biol Chem 1991; 266:5363-5366.

143. Blumer KJ, Thorner J. Receptor-G protein signalling in yeast. Annu Rev Physiol 1991; 53:37-57.

144. Tang W-J, Gilman AG. Type-specific regulation of adenylyl cyclase by G protein βγ subunits. Science 1991; 254:1500-1503.

145. Federman AD, Conklin BR, Schrader KA et al. Hormonal stimulation of adenylyl cyclase through Gᵢ-protein βγ subunits. Nature 1992; 356:159-161.

146. Camps M, Hou C, Sidiropoulos D et al. Stimulation of phospholipase C by guanine-nucleotide binding protein βγ subunits. Eur J Biochem 1992; 206:821-831.

147. Wu D, Katz A, Simon MI. Activation of phospholipase C β₂ by the α and βγ subunits of trimeric GTP-binding protein. Proc Natl Acad Sci USA 1993; 90:5297-5301.

148. Kleuss C, Hescheler J, Ewel C et al. Assignment of G-protein subtypes to specific receptors inducing inhibition of calcium currents. Nature 1991; 353:43-48.

149. Kleuss C, Scherubl H, Hescheler H et al. Different β subunits determine G-protein interaction with transmembrane receptors. Nature 1992; 358:424-426.

150. Kleuss C, Scherubl H, Hescheler J et al. Selectivity in signal transduction is determined by each of the subunits of heterotrimeric G-proteins. Science 1993; 259:832-834.

151. Thomas JM, Hoffman BB. Adenylate cyclase supersensitivity: a general means of cellular adaptation to inhibitory agonists? Trends Pharmacol Sci 1987; 8:308-311.

152. Milligan G. Agonist control of G protein levels. Trends Pharmacol Sci 1991; 12:207-209.

153. Hausdorff WP, Caron MG, Lefkowitz RJ. Turning off the signal: desensitization of β-adrenergic receptor function. FASEB J 1990; 4:2881-2889.

154. Wojcikiewicz RJH, Tobin AB, Nahorski SR. Desensitization of cell signalling mediated by phosphoinositidase C. Trends

Pharmacol Sci 1993; 14:279-285.

155. Sibley DR, Benovic JL, Caron MG et al. Regulation of transmembrane signalling by receptor phosphorylation. Cell 1987; 48:913-922.

156. Vicentini LM, DiVirgilio F, Pozzan T et al. Tumor promoter phorbol 12-myristate, 13-acetate inhibits phosphoinositide hydrolysis and cytosolic Ca$^{2+}$ rise induced by the activation of muscarinic receptors in PC12 cells. Biochem Biophys Res Commun 1985; 127:310-317.

157. Orellana SA, Solski PA, Brown JH. Phorbol ester inhibits phosphoinositide hydrolysis and calcium mobilization in cultured astrocytoma cells. J Biol Chem 1985; 260:5236-5239.

158. Hepler JR, Shelton H, Kendall T. Long-term phorbol ester treatment down-regulates protein kinase C and sensitizes the phosphoinositide signalling pathway to hormone and growth factor stimulation. J Biol Chem 1988; 263:7610-7619.

159. Thompson P, Findlay JB. Phosphorylation of ovine rhodopsin. Biochem J 1984; 220:773-780.

160. Inglese J, Glickman JF, Lorenz W et al. Isoprenylation of a protein kinase. J Biol Chem 1992; 267:1422-1425.

161. Wilden U, Hall SW, Kuhn H. Phosphodiesterase activation by photoexcited rhodopsin is quenched when rhodopsin is phosphorylated and binds the intrinsic 48-kDa protein of rod outer segments. Proc Natl Acad Sci USA 1986; 83:1174-1178.

162. Lefkowitz RJ. β-adrenergic receptor kinase: primary structure delineates a multigene family. Science 1989; 246:235-240.

163. Lefkowitz RJ. G protein-coupled receptor kinases. Cell 1993; 74:409-412.

164. Inglese J, Freedman NJ, Koch WJ et al. Structure and mechanism of the G protein-coupled receptor kinases. J Biol Chem 1993; 268:23735-23738.

165. Lohse MJ, Benovic JL, Codina J et al. β-arrestin: a protein that regulates β-adrenergic receptor function. Science 1990; 248:1547-1550.

166. Haga K, Haga T. Activation by G protein βγ subunits of agonist- or light-dependent phosphorylation of muscarinic acetylcholine

receptors and rhodopsin. J Biol Chem 1992; 267:2222-2227.

167. Kameyama K, Haga K, Haga T et al. Activation by G protein βγ subunits of β-adrenergic and muscarinic receptor kinase. J Biol Chem 1993; 268:7753-7758.

168. Pitcher JA, Inglese J, Higgins JB et al. Role of βγ subunits of G proteins in targeting the β-adrenergic receptor kinase to membrane-bound receptors. Science 1992; 257:1264-1267.

169. Kunapuli P, Benovic JL. Cloning and expression of GRK5: a member of the G protein-coupled receptor kinase family. Proc Natl Acad Sci USA 1993; 90:5588-5592.

170. Benovic JL, Gomez J. Molecular cloning and expression of GRK6. J Biol Chem 1993; 268:19521-19527.

171. Peralta EG, Ashkenazi A, Winslow JW et al. Differential regulation of PI hydrolysis and adenylyl cyclase by muscarinic receptor subtypes. Nature 1988; 334:434-437.

172. Ashkenazi A, Peralta EG, Winslow JW et al. Functionally distinct G proteins selectively couple different receptors to PI hydrolysis in the same cell. Cell 1989; 56:487-493.

173. Vallar L, Muca C, Magni M et al. Differential coupling of dopaminergic D$_2$ receptors expressed in different cell types. J Biol Chem 1990; 265:10320-10326.

174. Munshi R, Pang I-H, Sternweis PC et al. A$_1$ adenosine receptors of bovine brain couple to guanine nucleotide-binding proteins G$_{i1}$, Gi$_2$ and G$_o$. J Biol Chem 1991; 266:22285-22289.

175. Freissmuth M, Schutz W, Linder ME. Interactions of the bovine brain A$_1$-adenosine receptor with recombinant G protein α-subunits. Selectivity for rG$_i$α-3. J Biol Chem 1991; 266:17778-17783.

176. Berstein G, Blank JL, Smrcka AV et al. Reconstitution of agonist-stimulated phosphatidylinositol 4,5-bisphosphate hydrolysis using purified m1 muscarinic receptor, G$_{q/11}$, and phospholipase C-β1. J Biol Chem 1992; 267:8081-8088.

177. McClue SJ, Milligan G. Molecular interaction of the human α$_2$-C10-adrenergic receptor, when expressed in Rat-1 fibroblasts, with multiple pertussis toxin-sensitive gua-

nine nucleotide-binding proteins: studies with site-directed antisera. Mol Pharmacol 1991; 40:627-632.

178. Murray-Whelan R, Schlegel W. Brain somatostatin receptor G-protein interaction. J Biol Chem 1992; 267:2960-2965.

179. Simonds WF, Goldsmith PK, Codina J et al. $G_{i2}$ mediates $\alpha_2$-adrenergic inhibition of adenylate cyclase in platelet membranes: in situ identification with G$\alpha$ C-terminal antibodies. Proc Natl Acad Sci USA 1989; 86:7809-7813.

180. Gutowski S, Smrcka A, Nowak L et al. Antibodies to the $\alpha_q$ subfamily of guanine nucleotide-binding regulatory protein $\alpha$ subunits attenuate activation of phosphatidylinositol 4,5-biphosphate hydrolysis by hormones. J Biol Chem 1991; 266:20519-20524.

181. Offermanns S, Schultz G, Rosenthal W. Evidence for opioid receptor-mediated activation of the G-proteins, $G_o$ and $G_{i2}$ in membranes of neuroblastoma x glioma (NG-108) hybrid cells. J Biol Chem 1991; 266:3365-3368.

182. Schmidt A, Hescheler J, Offermanns S et al. Involvement of pertusis toxin-sensitive G-proteins in the hormonal inhibition of dihydropyridine-sensitive $Ca^{2+}$ currents in an insulin-secreting cell line (RINm5F). J Biol Chem 1991; 266:18025-18033.

183. Sho K, Okajima F, Majid MA et al. Reciprocal modulation of thyrotropin actions by $P_1$-purinergic agonists in FRTL-S thyroid cells. Inhibition of cAMP pathway and stimulation of phospholipase C-$Ca^{2+}$ pathway. J Biol Chem 1991; 266:12180-12184.

184. Biden TJ, Browne CL. Cross-talk between muscarinic and adenosine receptor signalling in the regulation of cytosolic free $Ca^{2+}$ and insulin secretion. Biochem J 1993; 293:721-728.

185. Katada T, Gilman AG, Watanabe Y et al. Protein kinase C phosphorylates the inhibitory guanine-nucleotide-binding regulatory component and apparently suppresses its function in hormonal inhibition of adenylate cyclase. Eur J Biochem 1985;

151:431-437.

186. Carlson KE, Brass LF, Manning DR. Thrombin and phorbol esters cause the selective phosphorylation of a G protein other than $G_i$ in human platelets. J Biol Chem 1989; 264:13298-13305.

187. Gollasch M, Kleuss C, Hescheler J et al. $G_{i2}$ and protein kinase C are required for thyrotropin-releasing hormone-induced stimulation of voltage-dependent $Ca^{2+}$ channels in rat pituitary GH$_3$ cells. Proc Natl Acad Sci USA 1993; 90:6265-6269.

188. Ferris CD, Huganir RI, Bredt DS et al. Inositol trisphosphate receptor: phosphorylation by protein kinase C and calcium calmodulin-dependent protein kinases in reconstituted lipid vesicles. Proc Natl Acad Sci USA 1991; 88:2232-2235.

189. Joseph SK, Ryan SV. Phosphorylation of the inositol trisphosphate receptor in isolated rat hepatocytes. J Biol Chem 1993; 268:23059-23065.

190. Biden TJ, Comte M, Cox JA et al. Calcium-calmodulin stimulates inositol 1,4,5-trisphosphate kinase activity from insulin secreting RINm5F cells. J Biol Chem 1987; 262:9437-9440.

191. Roach PJ. Multisite and hierarchal protein phosphorylation. J Biol Chem 1991; 266:14139-14142.

192. Sarndahl E, Bokoch GM, Stendahl O et al. Stimulus-induced dissociation of $\alpha$ subunits of heterotrimeric GTP-binding proteins from the cytoskeleton of human neutrophils. Proc Natl Acad Sci USA 1993; 90:6552-6556.

193. Ausiello DA, Stow JL, Cantiello HF et al. Purified epithelial Na$^+$ channel complex contains the pertussis toxin-sensitive $G\alpha_{i-3}$ protein. J Biol Chem 1992; 267:4759-4765.

194. Leiber D, Jasper JR, Alousi AA et al. Alteration in $G_s$-mediated signal transduction in S49 lymphoma cells treated with inhibitors of microtubules. J Biol Chem 1993; 268:3833-3837.

195. Strittmatter SM, Valenzuela D, Sudo Y et al. An intracellular guanine nucleotide release protein for $G_o$. J Biol Chem 1991; 266:22465-22471.

# STRUCTURAL DETERMINANTS OF RECEPTOR FUNCTION

Investigation of structure-function relationships in G protein-coupled receptors has been directed towards identification of residues and domains of the receptors which are of significance in ligand binding, functional activation, G protein coupling and desensitization of receptor activity. Much attention has been focused on rhodopsin, and the adrenergic and muscarinic acetylcholine receptor subtypes, since these were among the first members of the receptor superfamily to be isolated. Determination of structural determinants of function in these receptors has provided a valuable framework for delineation of structure-function relationships in other members of the receptor superfamily. It is becoming increasingly apparent that structural determinants of function are contributed by sequence features which are discontinuous within the primary protein sequence and which arise from the three-dimensional structure of the receptor protein and the molecular species with which it interacts, namely its extracellular ligand(s) and intracellular G proteins.

## LIGAND BINDING AND RECEPTOR ACTIVATION

A variety of chemical modification and photoaffinity labeling and detection methods have been used for mapping ligand interaction sites on rhodopsin and other receptors which interact with small ligands, such as the adrenergic and muscarinic acetylcholine receptors. In addition, since the isolation of receptor subtypes by molecular cloning, advantage has been taken of molecular biological approaches. Strategies employed for delineation of the structural determinants of ligand binding and functional activation of receptors have included site-directed mutagenesis, the construction of deletion mutants and chimeric molecules, the characterization of naturally occurring mutations which influence receptor function and the use of antibodies directed against receptor epitopes.

## 1. RHODOPSIN

The visual pigments, comprising rhodopsin and the color pigment proteins, differ from other members of the G protein-coupled receptor superfamily in that the ligand 11-*cis*-retinal is bound covalently to the receptor molecule and receptor activation occurs in response to absorption of photons of light. However, examination of ligand binding and receptor activation mechanisms in the visual pigment proteins has provided valuable information which is generally applicable to other members of the superfamily, such as the adrenergic and muscarinic acetylcholine receptors, which also interact with small ligands.[1]

Initial indications that covalent binding of the photoactivatable 11-*cis*-retinal moiety to rhodopsin involves transmembrane (TM) helices which constitute the hydrophobic core of the protein came from cross-linking and fluorescent energy transfer studies.[2,3] As depicted schematically in Figure 3.1a, the chromophore 11-*cis*-retinal is bound to rhodopsin via a protonated Schiff base linkage between the aldehyde group of retinal and the ε-amino group of Lys296, within TM helix VII. The Schiff base counterion is provided by Glu113, within TM helix III, which forms a salt bridge with the protonated retinal Schiff base at Lys296. The Schiff base counterion represents an important element in regulation of the wavelength of maximal absorption of the visual pigment.[4]

Photoexcitation involves isomerization of the chromophore, from 11-*cis*-retinal to all-*trans*-retinal (Fig. 3.1b), deprotonation of the Schiff base and disruption of the salt bridge.[5,6] Transition of the photoexcited rhodopsin to its active conformation, metarhodopsin II, is associated with a conformational change involving TM helix III of the rhodopsin protein[7,8] and is believed to involve protonation of His211, located within TM helix V.[9] The significance of the Schiff base linkage in maintaining the inactive conformation of rhodopsin is demonstrated by the identification of a mutant rhodopsin molecule which carries the amino acid substitution Lys296Glu. This mutant protein is constitutively active in the absence of both chromophore and light, and gives rise to one of the autosomal dominant forms of the degenerative disorder retinitis pigmentosa. Mutations of His211 are also observed in cases of autosomal dominant retinitis pigmentosa, although the mechanisms underlying the pathological phenotype remain to be defined (see chapter 4).

## 2. Adrenergic and Muscarinic Acetylcholine Receptor Subytpes

A large body of work has demonstrated that, as for rhodopsin, the binding of specific cationic biogenic amine ligands to adrenergic and muscarinic acetylcholine receptor subtypes occurs within the hydrophobic core of the receptor protein. The TM domains are necessary for ligand binding and confer ligand specificity, while the hydrophilic extracellular and intracellular domains are not directly involved in ligand binding.[10,11] However, the hydrophilic loops of the receptor play a role in receptor folding and conformation, and of particular significance in this regard are Cys residues occurring in extracellular loops which participate in disulfide bonding.[12,13] Formation of the ligand binding pocket is critically dependent upon correct protein folding and disposition of the receptor within the membrane, since TM domains which are not necessarily contiguous on the receptor sequence contribute structural determinants that mediate ligand binding and receptor activation.

It is beyond the scope of this review to discuss in detail the experimental data which have contributed to an extensive understanding of the ligand binding site of the adrenergic and muscarinic acetylcholine receptor subtypes. The current models for ligand binding interactions with the β$_2$-adrenergic and m3 muscarinic acetylcholine receptors will be presented. The reader is referred to a number of excellent review articles for a more detailed treatment.[12,14-22]

Endogenous agonists of the adrenergic receptor subtypes are catecholamines, which are characterized structurally by an aromatic catechol ring and a protonated amine group separated by a β-hydroxy ethyl chain. Specific molecular requirements for ligand binding to the adrenergic receptor subtypes include the provision of counterions for interaction with the protonated amine group, hydrogen bonding interactions with the β-hydroxyl and catechol hydroxyl groups, and aromatic interactions with the ring structure. The current view of the mode of interaction of the neurotransmitter

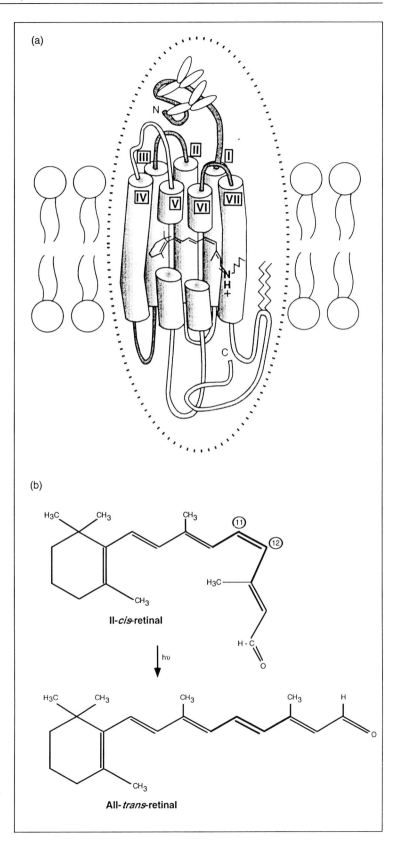

Fig. 3. 1. Binding of 11-cis-retinal to rhodopsin and photoisomerisation of the chromophore. (a) Model of rhodopsin. Transmembrane helices are numbered consecutively and sites of N-linked glycosylation within the N-terminal domain and palmitoylation within the C-terminal domain are indicated. The transected view through transmembrane helices V and VI shows binding of the chromophore 11-cis-retinal through a protonated Schiff base linkage to the side chain of the amino acid residue Lys296 within helix VII. [Reprinted with permission from FASEB Journal Volume 6, Hargrave PA and McDowell JH, Rhodopsin and phototransduction: a model system for G protein-linked receptors, Pages 2323-2331, Copyright (1992)]. (b) Light-catalyzed isomerization of 11-cis-retinal to all-trans-retinal.

noradrenaline (norepinephrine) with the $\beta_2$-adrenergic receptor is depicted in Figure 3.2a. The ligand is postulated to intercalate into the hydrophobic core of the receptor, and to have specific points of contact with amino acid residues on several of the TM helices. The counterion for the protonated amine group is provided by the carboxylate side chain of Asp113 in TM3. Hydrogen bonding between Ser165 in TM4 and the residues Ser204 and Ser207 in TM5 provide anchorage for the $\beta$-hydroxyl and catechol hydroxyl groups, respectively. Specific aromatic interactions with the catechol ring are believed to involve Phe289 and Phe290 in TM6 and Tyr326 in TM7.[18,23]

The importance of the structural integrity of the ligand binding pocket in receptor function is highlighted by the characteristics of a naturally occurring variant of the human $\beta_2$-adrenergic receptor, which carries the single amino acid substitution Thr164Ile in TM4. Mutation of this Thr residue located within the ligand binding pocket, adjacent to Ser165, results in a reduction in high affinity agonist binding and impairment of both functional coupling and agonist-stimulated receptor sequestration events.[24] These findings could reflect significant contributions of amino acid residues in this region to overall receptor conformation.

Structure-activity studies of adrenergic ligands have shown that the amine group and the $\beta$-hydroxyl substitutent are important for agonist and antagonist binding to the receptor. The catechol ring, on the other hand, is characteristic of agonists. Antagonists generally feature increased hydrophobicity of the aromatic ring structure and increased distance between the aromatic ring and the amine group by appropriate substitution of the aromatic ring.[25]

Receptor subtype selectivity for agonists and antagonists has been examined by the generation of chimeric receptors. The data obtained from such studies are consistent with agonist and antagonist structure-activity relationships and with the current model proposed for ligand interaction with

adrenergic receptors (see Fig. 3.2a). For example, it has been shown with chimeric $\beta_1/\beta_2$-adrenergic receptors that TM4 contributes to receptor subtype selectivity of agonist binding, while TM6 and 7 contribute to antagonist binding selectivity.[26] The involvement of TM7 in determining antagonist binding specificity has also been shown with chimeric $\alpha_{2A}/\beta_2$-adrenergic receptors. In addition, substitution of Phe412 in TM7 of the $\alpha_{2A}$-adrenergic receptor with Asn, which occurs in the corresponding position in the $\beta_1$- and $\beta_2$-adrenergic receptors, results in loss of binding of the $\alpha_{2A}$-adrenergic receptor antagonist yohimbine and acquisition of high affinity binding for a particular class of $\beta_1$- and $\beta_2$-adrenergic receptor antagonists.[27,28] In contrast, a proteolytic product of the porcine $\alpha_2$-adrenergic receptor which contains only the first five TM domains of the receptor has been shown to retain antagonist binding activity. It is possible that TM7 may be important in directing the folding of the ligand binding pocket, but may be dispensable for ligand binding to the receptor once this pocket has been formed.[29]

A number of amino acid residues which are involved in receptor activation and signal transmission have been identified. These include Asp79, which occurs in TM2, Asp130 at the intracellular face of TM3, and Asn312 and Tyr316 in TM7. The residues Asp79 and Asp130 are conserved among most members of the G protein-coupled receptor superfamily (see chapter 1, Fig. 1.5), suggesting that some common features may underlie mechanisms of activation of many receptors in the superfamily. A multistep dynamic model for activation of the $\beta_2$-adrenegic receptor has been proposed, whereby agonist-induced interaction of Asp79 with Tyr316 (see Fig. 3.2a) may constitute the first step towards G protein activation.[30]

The endogenous ligand for muscarinic acetylcholine receptor subtypes is the cationic biogenic amine acetylcholine. All muscarinic acetylcholine receptor subtypes contain an Asp residue in TM3 which corresponds to Asp113 of the $\beta_2$-adrenergic

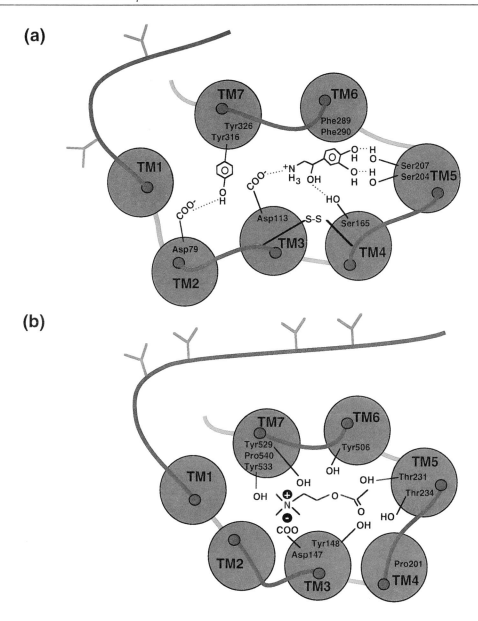

*Fig. 3. 2. Ligand binding to adrenergic and muscarinic acetylcholine receptors. Schematic view of the interaction of (a) noradrenaline with amino acid residues of the β₂-adrenergic receptor, and (b) acetylcholine with amino acid residues of the rat m3 muscarinic acetylcholine receptor. Transmembrane helices are depicted as viewed from the outside of the cell, with heavy lines connecting helices representing extracellular domains of the receptor and pale lines representing intracellular domains. Disulfide bonding between the first and second extracellular loops is defined (—S-S—) and sites for N-linked glycosylation within the N-terminal extracellular domain are shown (Y). Amino acid residues of importance in ligand binding or receptor activation are identified. [(a) Reprinted with permission from Protein Science Volume 2, Strosberg AD, Structure, function, and regulation of adrenergic receptors, Pages 1198-1209, Copyright (1993). (b) Reprinted with the permission of Cambridge University Press from Life Sciences Volume 53, Wess J, Mutational analysis of muscarinic acetylcholine receptors: structural basis of ligand/receptor/G protein interactions, Pages 1447-1463, Copyright (1993), with kind permission from Elsevier Science Ltd. The Boulevard, Langford Lane, Kidlington OX5 1GB, UK].*

receptor. Biochemical and molecular biological approaches have been used to establish the involvement of this residue in interaction with the cationic head group of muscarinic acetylcholine receptor agonists and antagonists, in a manner analogous to the interaction of Asp113 of the β₂-adrenergic receptor with the protonated amine group of catecholamines.[17,19,22] Other receptors which bind cationic biogenic amine ligands, including the dopamine and serotonin receptor subtypes, also exhibit conservation of an Asp residue in TM3,[31] suggesting a critical role for this residue in interaction with the cationic head group of the respective ligands of these receptors also.

Among other amino acid residues conserved in TM domains of muscarinic acetylcholine receptor subtypes are a number of Ser, Thr and Tyr residues.[31] Site-directed mutagenesis studies of the rat m3 muscarinic acetylcholine receptor have established contributions of Tyr148 in TM3, Thr231 and Thr234 in TM5, Tyr506 in TM6, and Tyr529 and Tyr533 in TM7, to the acetylcholine binding domain of this receptor subtype.[32] The side chains of these residues are believed to interact with one or more electron-rich centers characteristic of muscarinic agonists, such as the ester moiety of acetylcholine or the carbamoyl group of carbachol. The residue Pro201 in TM4, which corresponds to a residue conserved in most G protein-coupled receptors (see chapter 1, Fig. 1.5), is believed to function in stabilization of receptor conformation for high affinity binding of agonists and antagonists.[33] The current model for interaction of acetylcholine with the rat m3 muscarinic acetylcholine receptor is shown in Figure 3.2b.

The conserved amino acid residues Thr231 and Thr234 in TM5 of the m3 muscarinic acetylcholine receptor, which contribute to the ligand binding site, correspond to conserved Ser204 and Ser207 residues of the β₂-adrenergic receptor, which are also involved in interaction with ligand (see Fig. 3.2a). Conservation of one Thr residue in TM5 is also seen in serotonin

receptor subtypes, while two corresponding Ser residues are conserved in TM5 of dopamine receptor subtypes.[31] The conservation of discrete amino acid residues in a number of receptors which interact with structurally related ligands suggests similarity in the binding interactions mediating ligand recognition. It has been postulated that the binding of all biogenic amine ligands occurs in a similar plane within the hydrophobic receptor core. This would lie approximately two helical turns away from the extracellular surface of the plasma membrane, at a distance of approximately 10-15Å from the membrane surface.[22]

The construction of chimeric muscarinic acetylcholine receptors has been used to investigate receptor subtype selectivity for agonists and antagonists. From studies with chimeric human m2/rat m3 receptors, the binding affinity for the receptor agonist carbachol has been demonstrated to depend upon the third intracellular loop, presumably by indirect effects on overall receptor conformation.[34] The ligand binding properties of chimeric m1/m2, m2/m3 and m2/m5 receptor constructs, on the other hand, have established that receptor subtype selectivity for receptor antagonists cannot be ascribed to any one discrete receptor segment. Antagonist binding selectivity involves a number of different TM and extracellular domains of the receptors, with different classes of antagonist discriminating different structural determinants.[20,35-37] Further studies incorporating site-specific mutagenesis approaches based on comparison of receptor subtypes sequences may provide indications of specific sequence features which contribute to antagonist binding selectivity.[38]

As observed for adrenergic receptor subtypes, a pivotal role in mediation of conformational changes associated with receptor activation may be assigned to the Asp113 residue in TM2 of the rat m3 muscarinic acetylcholine receptor. This residue corresponds to Asp 79 in the β₂-adrenergic receptor (see Fig. 3.2a)[33] and is conserved among most members of the receptor superfamily (see chapter 1, Fig. 1.5).

Studies incorporating site-directed mutagenesis have identified a role for the residues Thr234 in TM5 and Tyr506 in TM6 not only in ligand binding, but also in functional activation of the rat m3 muscarinic acetylcholine receptor. It is believed that these two residues mediate acetylcholine-induced conformational changes in TM5 and 6 to the connecting third intracellular loop of the receptor, which in turn results in the activation of specific G proteins.[39] Similarly, Pro540 of this receptor, which occurs in TM7 and corresponds to a Pro residue conserved among most G protein-coupled receptors (see chapter 1, Fig. 1.5), has been shown to be important for agonist-induced conformational activation of the receptor protein.[33]

## 3. METABOTROPIC GLUTAMATE RECEPTORS

The metabotropic glutamate receptors (mGluR)1-7 show a common structural architecture with other members of the G protein-coupled receptor superfamily, comprising an extracellular N-terminal domain that precedes 7 TM segments and an intracellular C-terminal domain.[40-44] In contrast to other G protein-coupled receptors which bind small ligands, such as the adrenergic and muscarinic acetylcholine receptors, the mGluRs are characterized by an unusually large N-terminal extracellular domain, ranging in size from 550 to 575 amino acid residues. In addition, the mGluR subtypes are closely related to one another in primary sequence but share very little primary sequence similarity to the majority of G protein-coupled receptors (see chapter 1, Figures 1.5 and 1.6). Within the N-terminal extracellular domain, the mGluRs exhibit a striking conservation of Cys residues, which presumably reflects an important role for these residues in formation or maintenance of the structural integrity of these receptors.

According to sequence homology, signal transduction properties and agonist selectivities, the seven mGluR subtypes may be subdivided into three subgroups, comprising mGluR1 and 5, mGluR2 and 3,

and mGluR4, 6 and 7. The role of the N-terminal extracellular domain in ligand binding properties of mGluR subtypes has been investigated by the construction of a series of chimeric mGluR1/mGluR2 molecules. Replacement of almost the entire N-terminal extracellular domain of mGluR1 with the corresponding region of mGluR2 results in a chimeric molecule with agonist selectivity characteristic of the mGluR2 subtype. Agonist selectivity has further been shown to reside within the N-terminal half of this extracellular domain.[45] Assignment of agonist selectivity to more discrete domains within the extracellular segment of these receptors has not been possible due to adverse effects of chimeric construction on receptor function, possibly arising from inappropriate incorporation of chimeric receptors into the membrane.

The available data are consistent with the N-terminal extracellular domain of these receptors representing the ligand binding site. The binding of agonist may induce a conformational change which is transmitted across the plasma membrane, resulting in activation of intracellular G proteins. However, more complex models, incorporating an involvement of the N-terminal extracellular and TM domains in ligand recognition and receptor activation, have also been proposed,[45] and definition of the mechanism of activation of mGluRs clearly constitutes a subject for further mutational analyses.

## 4. GLYCOPROTEIN HORMONE RECEPTORS

The N-terminal extracellular domain of glycoprotein hormone receptors comprises 340-400 residues and contains Cys-rich regions that are believed to play a role in maintaining the structural integrity of this domain. The receptors for the hormones follicle-stimulating hormone (FSH), luteinizing hormone/chorionic gonadotropin (LH/CG) and thyroid-stimulating hormone (TSH) also contain between 9 and 14 copies of an imperfectly repeated "leucine-rich repeat" sequence of approximately 25 amino acids, which is characteristic of leucine-rich glycoproteins (LRG; see chapter 1).

The construction of chimeric receptors between members of the glycoprotein hormone receptor family and other members of the G protein-coupled receptor superfamily has established that the N-terminal extracellular domain has ligand recognition properties and is the critical determinant for ligand binding specificity. For example, a chimeric receptor generated by the fusion of the N-terminal extracellular domain of the rat LH/CG receptor with the TM and C-terminal domains of the hamster $\beta_2$-adrenergic receptor exhibits high affinity binding of both human choriogonadotropin (hCG) and the β-adrenergic antagonist cyanopindolol.[46] These data suggest that the N-terminal extracellular domain of LH/CG receptor is both necessary and sufficient for high affinity binding of ligand. They also confirm localization of the ligand binding site in adrenergic receptors to the TM domains, as discussed above (Section 2). High affinity binding of hCG to the isolated N-terminal extracellular domain of LH/CG receptor, expressed as a soluble protein, is consistent with these findings.[47,48]

The glycoprotein hormones are large heterodimeric proteins of molecular mass 28-38 kilodaltons (kD). They comprise an α subunit which is common to all three hormones, and a β subunit which is responsible for biological specificity.[49] A combination of approaches, including deletion mutagenesis, construction of chimeric receptors, competition with synthetic peptides and the use of antibodies which recognize receptor epitopes, has established that within the N-terminal extracellular domain of the glycoprotein hormone receptors, there are many points of contact of receptor with ligand. Discontinuous segments of the extracellular domain contribute to the ligand binding interactions, highlighting the importance of receptor conformation for generation of the ligand binding site.

Using truncation mutants of the rat LH/CG receptor, it has been shown that only the first 8 of 14 leucine-rich repeat

sequence motifs, comprising amino acid residues 1-206, are required for high affinity binding of hCG. Construction of a chimeric receptor, with amino acid residues 1-146 of the LH/CG receptor replacing the corresponding segment of the FSH receptor, results in a chimeric molecule with hCG binding specificity. The converse experiment, with introduction of N-terminal FSH receptor sequence to replace the N-terminal domain of the LH/CG receptor, identified a requirement for 11 of 14 leucine-rich repeat sequence motifs, comprising amino acids 1-283, to confer high affinity FSH binding on the chimeric receptor.[50-52] A specific involvement of rat FSH receptor residues 9-30 in ligand binding has been reported.[53] Similar studies addressing sites of ligand interaction with the human receptors would be of interest, particularly in light of the ligand specificity of the human LH/CG receptor. The human LH/CG receptor recognizes human LH and hCG, but not rat LH, while the rat LH/CG receptor binds both rat LH and hCG with high affinity.[51]

Contributions to the ligand binding site of the TSH receptor arise from receptor segments within the early N-terminal, mid-region and membrane-proximal region of the extracellular domain. Chimeric human TSH/rat LH/CG receptors substituted for the amino acid residues 1-82 by the corresponding segment of the LH/CG receptor do not bind TSH.[54] Within this region, hormone binding activity is associated in particular with amino acid residues 12-50.[55-58] Additional regions of importance for ligand binding encompass residues 106-125 in the N-terminal region, and residues 171-260 in the mid-region of the extracellular domain of the human TSH receptor,[54,58,59] with a particular involvement of residues 201-211 and 222-245.[51,58,60] In the membrane-proximal region, the properties of deletion mutants of the rat TSH receptor initially identified the amino acid residues 299-301 and 387-395 as potential hormone binding sites. Peptide competition assays have confirmed ligand interaction with residues 286-305, and identified an additional contact site encompassing residues 256-275. The use of site-directed mutagenesis has established a role for Tyr385, Thr388 and Asp403 in ligand binding, and demonstrated the importance of Cys41, Cys301, Cys390 and Cys398, presumably by involvement in intramolecular disulfide bond formation and contribution to the overall conformation of the N-terminal extracellular domain.[58,61,62]

The autoimmune diseases hypothyroid idiopathic myxedema and hyperthyroid Graves' disease arise from the generation of autoantibodies directed against the TSH receptor. The autoantibodies responsible for the respective disease phenotypes recognize discrete epitopes within the N-terminal extracellular domain of the TSH receptor sequence. The relationship of these epitopes to regions of the receptor which are of significance in ligand binding interactions will be discussed in detail in chapter 4.

Peptide competition experiments have been used to identify specificity determinants that provide for discrimination between the glycoprotein hormone ligands by their respective receptors. Peptides corresponding to residues 16-35, 106-125, 226-245, 256-275 and 286-305 of the N-terminal extracellular domain of the human TSH receptor represent sites of ligand interaction and will inhibit the binding of TSH to native TSH receptors. However, specificity for interaction with TSH is a property only of peptides corresponding to residues 226-245 and 286-305. Peptides corresponding to residues 16-35, 106-125 and 256-275 of TSH receptor sequence are also able to inhibit the binding of hCG to native LH/CG receptors, consistent with identification of corresponding regions of the human LH/CG receptor, comprising amino acid residues 21-38, 102-115 and 253-272, as potential hCG contact sites.[58,63] These sites would presumably be involved in recognition of a structural determinant common to glycoprotein hormones, such as the α subunit. Of these three sites, TSH receptor sequences 106-125 and 256-275 are largely homologous to the corresponding LH/CG receptor sequences, while the

two TSH receptor sites that confer ligand binding specificity are largely nonhomologous to corresponding sequences of the LH/CG receptor. Similarly, the segment comprising amino acid residues 9-30 within the N-terminal extracellular domain of the FSH receptor, which has no sequence homology with the corresponding regions of LH/CG or TSH receptors, has FSH recognition properties.[53,64] and has been demonstrated to interact with the hormone-specific β subunit of FSH. In addition, it has recently been shown that distinct negative determinants within respective N-terminal domains of the glycoprotein hormone receptors, which restrict ligand-receptor interaction, contribute to the ligand binding specificity of these receptors.[65]

Peptide competition experiments conducted with the human LH/CG receptor have also indicated involvement of the third extracellular loop, corresponding to residues 573-583, in ligand binding interactions.[63] In addition, a potential role for TM domains in binding of glycoprotein hormone ligands has been suggested from examination of the binding and activation properties of LH/CG receptor constructs in which virtually the entire N-terminal extracellular domain has been deleted.[66] It was found that hCG can bind to these truncated receptors with low affinity, and cause functional activation. Subsequently, it has been reported that in full-length rat LH/CG receptors, the residue Asp383 in TM2, which corresponds to an Asp residue highly conserved in TM2 of G protein-coupled receptors (see chapter 1, Fig. 1.5), plays a role in high affinity hormone binding and resultant receptor activation.[67] However, conflicting data, suggesting involvement of this residue in allosteric regulation of the binding affinity of LH/CG receptor in response to sodium ions, have also been obtained.[68] These data are consistent with the role of this conserved Asp residue in allosteric regulation by sodium ions of agonist binding to $\alpha_2$-adrenergic receptors.[52,69]

The mechanism whereby ligand binding triggers activation of glycoprotein hormone receptors remains to be defined. There is evidence that receptor activation and signal generation are events distinct from high affinity hormone binding. The residue Asp397, which is located at the extracellular face of TM2 of the LH/CG receptor, and is uniquely conserved in glycoprotein hormone receptors, has been identified as being important for signal generation but not for hormone binding. Receptor activation has been shown to involve specific interaction of Asp397 of the receptor with Lys91 on the α-subunit of hCG.[70,71]

## 5. PEPTIDE RECEPTORS

Efforts to identify sites of ligand interaction for receptors which bind bioactive peptides have involved site-directed mutagenesis of discrete receptor subtypes and the construction of chimeric receptors between receptor subtypes which interact with the same or closely related ligands. An increasing availability of peptidergic and nonpeptidergic agonists and antagonists is facilitating such analyses. Ligand recognition determinants are beginning to be defined for a range of receptors, including the angiotensin II, δ-opioid, N-formyl peptide and bombesin/neuromedin B receptors,[72-77] the endothelin receptor subtypes $ET_A$ and $ET_B$,[78-80] and receptors for the mammalian tachykinin peptides substance P, neurokinin A (NKA; substance K) and neurokinin B (NKB; neuromedin K).

The tachykinin peptide receptor subtypes, designated $NK_1$, $NK_2$ and $NK_3$, bind with high affinity the peptides substance P, NKA and NKB, respectively. However, all three receptors exhibit a characteristic rank order of affinity for the binding of all three tachykinin peptides. This reflects recognition of the characteristic C-terminal sequence, (-Phe-X-Gly-Leu-Met-NH$_2$), which is common to all three peptides and is required for biological activity of the peptides. The N-terminal sequence differs between the peptides and may be the primary determinant of receptor subtype specificity (Fig. 3.3a).[81]

Examination of the ligand binding properties of wild-type receptors, chimeric receptors constructed between $NK_1$ and $NK_2$, and $NK_1$ and $NK_3$ receptors, as well as receptors carrying site-specific mutations or deletions, has established that extracellular and TM domains of the receptors contribute to recognition and binding of endogenous tachykinin ligands.

Optimal binding of all three peptides to the human $NK_1$ receptor has been shown to rely critically upon discrete residues within the N-terminal, the first and second extracellular domains and TM domains. These include Asn23, Gln24 and Phe25 in the N-terminal domain, Asn96 and His108 in the first extracellular loop, residues 176-183 in the second extracellular loop, as well as Glu78, Asn85, Asn89, Tyr92 and Asn96 in TM2, Tyr205 in TM5 and Tyr287 in TM7 (Fig. 3.3b).[82] Of particular interest also is the residue Glu97 in the first extracellular loop, which plays a role in interaction of the human $NK_1$ receptor with NKB, but not with substance P or NKA. The available data have identified both overlapping and distinct functional determinants for binding of discrete ligands to the $NK_1$ receptor. With respect to the other receptor subtypes, a number of the residues of importance for ligand binding interactions of the $NK_1$ receptor are not conserved in the $NK_2$ or $NK_3$ receptor subtypes (Fig. 3.3b), suggesting lack of conservation of peptide binding sites across the three receptor subtypes. On the other hand, the residue corresponding to Glu97 in the first extracellular loop of the human $NK_1$ receptor is conserved in the human $NK_3$ receptor, consistent with it playing a role in the binding of NKB to the $NK_3$ receptor also.[83-85]

Receptor subtype specificity for high affinity ligand binding and activation by substance P appears to involve segments of the $NK_1$ receptor encompassing the N-terminal extracellular domain and including the first and second extracellular loops.[86,87] In contrast, specific recognition of NKA by the $NK_2$ receptor involves structural determinants located in the segment of the receptor extending from the end of TM5 to within TM7, and including the third

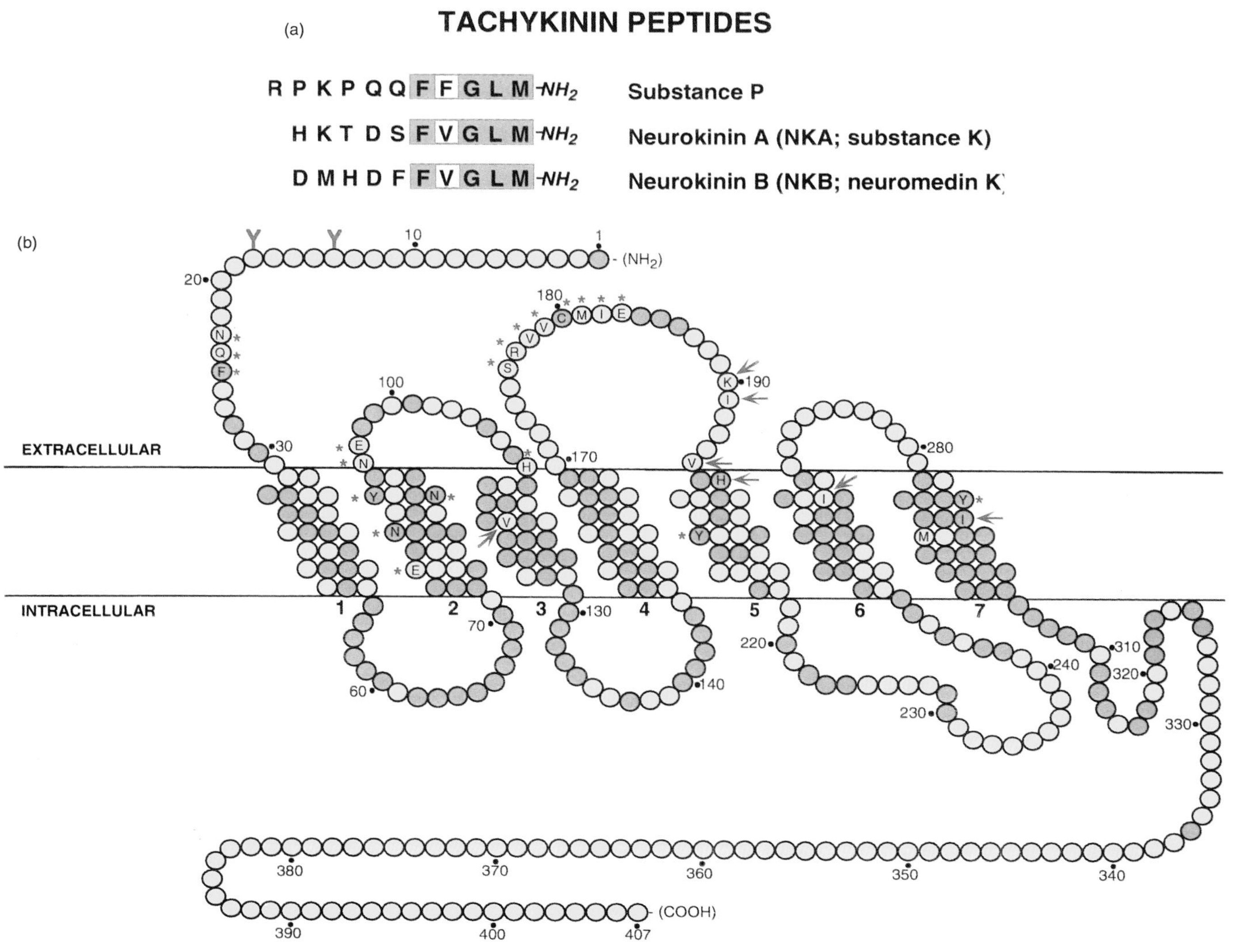

## TACHYKININ PEPTIDES

(a)

R P K P Q Q F F G L M -$NH_2$    Substance P

H K T D S F V G L M -$NH_2$    Neurokinin A (NKA; substance K)

D M H D F F V G L M -$NH_2$    Neurokinin B (NKB; neuromedin K)

(b)

EXTRACELLULAR

INTRACELLULAR

Fig. 3.3. Tachykinin peptide ligands and ligand binding interactions with tachykinin peptide receptors (a) Amino acid sequence of the mammalian tachykinin peptides substance P, neurokinin A and neurokinin B. (b) Model of the human NK1 receptor. Transmembrane helices are numbered consecutively and potential sites for N-linked glycosylation are identified (Y). Residues conserved among NK1, NK2, and NK3 receptors are shown in blue. Discrete residues identified to date to be of importance in agonist (*) and antagonist (→) binding interactions of the NK1 receptor are identified.

extracellular loop.[88] Similarly, significant determinants for NKB binding specificity appear to be localized in the C-terminal part of the NK₃ receptor, extending from within the second extracellular loop and incorporating the third extracellular loop, as well as TM7.[83,85,87,89] The substitution of residues 170-174 in the second extracellular loop and residues 271-280 in the third extracellular loop of the human NK₁ receptor with the corresponding sequence of the human NK₃ receptor[83] results in increased affinity for NKB without affecting affinity for substance P. A role for TM7 in NKB binding has been demonstrated by site-directed mutagenesis of the human NK₁ receptor to generate a receptor carrying the mutation Met291Phe within TM7. The residue Phe occurs within the corresponding position of the human NK₃ receptor, and its introduction into the framework of the NK₁ receptor results in increased affinity of the mutant NK₁ receptor for NKB, with no effect on substance P binding.[83]

It has been demonstrated that ligand binding of agonists and antagonists to the NK₁ receptor involves distinct and different structural determinants. In contrast to determinants for the binding of substance P, binding of the potent, selective, nonpeptidergic antagonist CP-96,345 to the human NK₁ receptor involves epitopes within TM5 and 6 which flank the second and third extracellular loops, respectively.[89] Of significance also is the specific residue Ile290 in TM7, with an additional involvement of Val116 in TM3, Lys190, Ile191 and Val195 in the second extracellular loop, His197 in TM5 and Ile266 in TM6.[90-92] The identification of specific molecular determinants for interaction with antagonists at the NK₁ and other peptide receptors should provide the basis for rational design of more potent and specific antagonist molecules.

## 6. THROMBIN RECEPTOR

Activation of the thrombin receptor involves a novel proteolytic mechanism, whereby thrombin, which is a multifunc-

tional protease, cleaves its receptor to create a new amino terminus that functions as a tethered ligand to activate the receptor.

Substrate cleavage by thrombin occurs through recognition across an extended substrate binding surface of residues both amino and carboxyl to the substrate cleavage site. Thrombin cleavage recognition determinants, which occur within the N-terminal extracellular domain of the thrombin receptor, are shown in Figure 3.4a. The first of these corresponds to amino acid residues 38-42. It comprises the sequence (Leu-Asp-Pro-Arg-Ser), which resembles the known thrombin cleavage site (Leu-Asp-Pro-Arg-Ile) in the thrombin-activated zymogen protein C. The second recognition determinant spans residues 51-60 of the human thrombin receptor and residues 52-62 of the murine thrombin receptor and has been denoted a hirudin-like domain, due to homology with the carboxyl tail of hirudin. This is a leech-derived anticoagulant that binds thrombin avidly and interacts with the anion-binding exosite of thrombin (Fig. 3.4b).[93,94]

Studies with mutant thrombin receptors and receptor peptides have shown that cleavage of the receptor protein, which occurs after Arg41, is necessary and sufficient for receptor activation by thrombin. Mutation of the cleavage site such that it could not be cleaved prevents receptor activation.[95] Synthetic peptides mimicking the cleavage recognition determinants of the receptor are cleaved by thrombin,[93] while uncleavable "mutant" synthetic peptides mimicking this region will inhibit the activity of thrombin against its receptor and other substrates.[93] In addition, replacement of the thrombin cleavage recognition sequence (Leu-Asp-Pro-Arg-Ser) with that for the protease enterokinase (Asp-Asp-Asp-Asp-Lys) results in altered receptor specificity, with cells expressing this construct responding to enterokinase and not to thrombin.[93,96] Recent kinetic studies have shown correlation of receptor cleavage with activation of signaling pathways.[97]

Proteolytic cleavage of receptor protein by thrombin results in the creation of a new N-terminus which acts as a tethered ligand for receptor activation (Fig. 3.4c). Synthetic peptides corresponding to the new N-terminal sequence are full agonists for receptor activation, with the first five amino acids of the receptor's agonist peptide domain (Ser-Phe-Leu-Leu-Arg) being sufficient to specify agonist activity[95,98-100] and bypass the requirement for receptor proteolysis. The protonated amino group of Ser1, which is unmasked only by cleavage of the receptor, and the phenyl ring of Phe2, have been shown to be critical for agonist activity.[99-101] However, subsequent events in receptor activation have not yet been elucidated.

## 7. OTHER RECEPTORS

Ligand binding sites are beginning to be defined for a variety of other receptors. For example, functional determinants for ligand binding to the interleukin (IL)-8 A receptor have been shown to occur within the N-terminal extracellular domain and the second extracellular loop. The IL-8 A and B receptor subtypes exhibit differential ligand binding properties, with the IL-8 A receptor binding IL-8 exclusively and the IL-8 B receptor binding both IL-8 and melanoma growth stimulatory activity (MGSA/GRO) with high affinity and neutrophil activating protein (NAP)-2 with low affinity. These differences are attributable to sequence heterogeneity within the N-terminal extracellular domain of the two receptor subtypes.[102-104]

The large number of odorant receptor sequences described to date[105-109] exhibit greatest sequence divergence in TM3, 4 and 5. This variability in sequence may reflect divergence in the ligand binding domain of these receptors, with individual receptors being capable of recognizing only one or a small set of structurally related odorant molecules.

## POST-TRANSLATIONAL MODIFICATIONS

Three forms of post-translational modification effected during biosynthesis of G protein-coupled receptors have been described.

The first of these is Asn-linked (*N*-linked) glycosylation, which results in the covalent attachment of carbohydrate residues to extracellular domains of the protein. The second form of modification involves linkage of the long chain saturated fatty acid moiety palmitate to a Cys residue in the C-terminal cytoplasmic domain. This results in attachment of the receptor to the intracellular face of the plasma membrane. Conservation of a Cys residue in the intracellular C-terminal cytoplasmic tail of the majority of G protein-coupled receptors (see chapter 1, Fig. 1.5) suggests palmitoylation to be a feature of most members of the superfamily. Cys residues in the first and second extracellular loops are believed to participate in disulfide bond formation, a third type of post-translational modification which is crucial for the function of a number of receptors.

A further form of post-translational modification is seen in phosphorylation of particular residues located in intracellular domains of receptors. Receptor phosphorylation is of fundamental significance in desensitization processes and is discussed in detail in the last section of this chapter.

## 1. GLYCOSYLATION

Sites for *N*-linked glycosylation occur in the extracellular N-terminal domain and/or first or second extracellular loops of the majority of G protein-coupled receptors (see chapter 1). The attachment of carbohydrate moieties is predicted to occur at Asn residues at the consensus sequence (Asn-X-Ser/Thr), where X is any amino acid residue apart from Pro or Asp.[110]

Site-specific mutagenesis of both the $\beta_2$-adrenergic and m2 muscarinic acetylcholine receptors has established that *N*-linked glycosylation is not essential for ligand binding or functional activation of either of these receptor proteins.[111,112] These data are consistent with earlier work, where deglycosylation of $A_1$ adenosine, β-adrenergic or muscarinic acetylcholine receptors was achieved using a variety of enzymatic or chemical treatments. No adverse effect on ligand binding or functional receptor

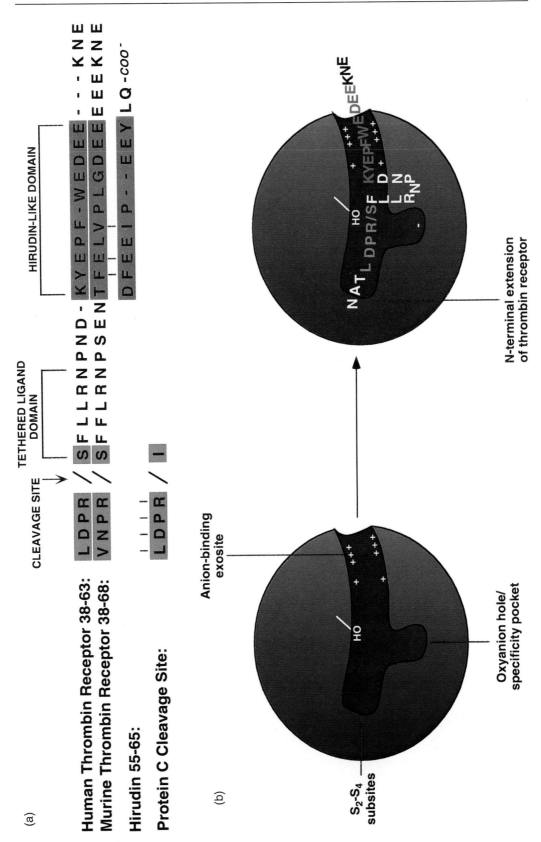

(a)

CLEAVAGE SITE

TETHERED LIGAND DOMAIN

HIRUDIN-LIKE DOMAIN

Human Thrombin Receptor 38–63:   LDPR / SFLLRNPND - KYEPF - WEDEE - - - KNE
Murine Thrombin Receptor 38–68:  VNPR / SFFLRNPSENTFELVPLGDEE EEKNE

Hirudin 55–65:                   DFEEIP - - EEY LQ -COO⁻

Protein C Cleavage Site:         LDPR / I

(b)

Anion-binding exosite

Oxyanion hole/ specificity pocket

S₂–S₄ subsites

N-terminal extension of thrombin receptor

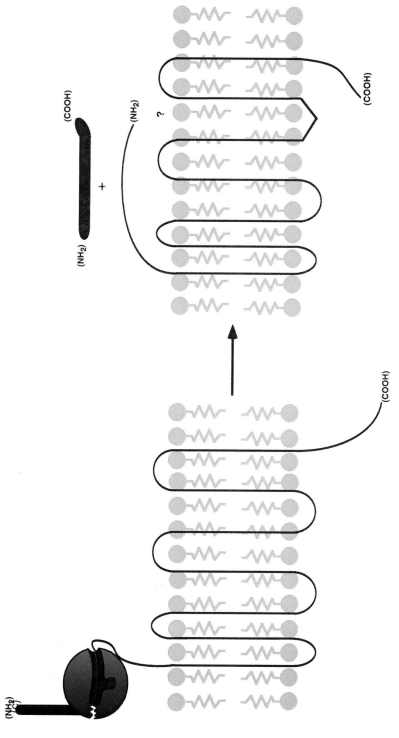

*Fig. 3.4. Activation of the thrombin receptor. (a) Thrombin cleavage recognition determinants within the N-terminal domain of the thrombin receptor. The thrombin cleavage recognition sequence and cleavage site, tethered ligand domain and hirudin-like domain, which corresponds to the domain of the receptor that binds to the anion-binding exosite of thrombin, are identified. The sequence of the human thrombin receptor is aligned with that of the murine thrombin receptor and these sequences are compared to the anion-binding exosite binding sequence of the leech-derived anticoagulant hirudin and the cleavage site for thrombin in the zymogen protein C. (b) Model of the interaction of thrombin receptor domains identified in (a) with thrombin. The extended substrate binding surface of thrombin is represented as a depression extending laterally across the surface of the protein. This region recognizes amino acid residues N- and C-terminal to the cleavage site in the substrate. The cleavage site interacts with the S1-S4 subsites of thrombin, while the hirudin-like domain of the receptor interacts with the anion-binding exosite. (c) Representation of proteolytic cleavage of thrombin receptor by thrombin, leading to exposure of a new N-terminus that functions as a tethered receptor agonist. [(a) Reprinted with permission from Thrombosis and Haemostasis Volume 70, Coughlin SR, Thrombin receptor structure and function, Pages 184-197, Copyright (1993). (b) and (c) Reprinted with permission from Nature Volume 353, Vu T-K, Wheaton VI, Hung DT et al, Domains specifying thrombin-receptor interaction, Pages 674-677, Copyright (1991) Macmillan Magazines Limited].*

coupling was observed, nor could a role for glycosylation in enhancement of receptor stability or protection from proteolysis be identified.[113-115] However, glycosylation-deficient mutant $\beta_2$-adrenergic receptors do exhibit altered subcellular distribution, with a significant reduction in the amount of receptor appearing at the cell surface. This suggests a role for N-glycosylation in intracellular trafficking of the $\beta_2$-adrenergic receptor. Since no adverse effect on cellular localization was observed with similarly mutated m2 muscarinic acetylcholine receptors, it is not possible to ascribe a general role for glycosylation in membrane targeting of G protein-coupled receptors. However, it is of interest to note that destruction of one of the two sites for N-linked glycosylation within the N-terminal intradiscal domain of the rhodopsin protein, with the naturally-occurring mutation Thr17Met, underlies one form of autosomal dominant retinitis pigmentosa, a degenerative disease of the human retina (see chapter 4).

Receptors for the glycoprotein hormones FSH, LH/CG and TSH contain up to six potential N-linked glycosylation sites within the N-terminal extracellular domain. It has been established that the glycoprotein hormone receptors are glycosylated proteins. While it is not known which of the potential glycosylation sites within the N-terminal domain are functional, the use of enzymatic means of carbohydrate removal and site-specific mutagenesis studies of the LH/CG receptor have generally established a lack of contribution of carbohydrate moieties to ligand binding and signal transduction events. However, as for the $\beta_2$-adrenergic receptor, a role has been suggested for one of the residues of the LH/CG receptor, which represents a potential glycosylation site (Asn173), in the localization of receptors to the plasma membrane.[52,116] Abolition of high affinity ligand binding and functional coupling in site-specific mutants of the TSH receptor may reflect a similar role for particular Asn residues (Asn77, Asn113) in expression of functional TSH receptors.[51,117]

The available data do not provide a clear picture of the contribution of glycosylation to functional expression of receptors for peptide ligands. For example, a lack of effect of the absence of glycoslyation on ligand binding has been reported for receptors for the peptides angiotensin II, substance P and N-formyl peptide.[72,84,118] In addition, high affinity ligand binding is observed when endothelin $ET_B$, neuropeptide Y (NPY) Y1 and neurotensin receptors are expressed in bacterial systems, where N-linked glycosylation does not occur.[119-121] On the other hand, carbohydrate moieties do appear to play a role in high affinity binding of agonists by native cholecystokinin (CCK), somatostatin and vasoactive intestinal peptide (VIP) receptors.[122-124] Similarly, the functional significance of lack of glycosylation in the subset of G protein-coupled receptors which do not carry potential glycosylation sites remains to be determined. For example, human platelet-activating factor (PAF) receptor contains no potential N-linked glycosylation sites within the N-terminal extracellular domain, while its counterpart expressed in the guinea pig contains one such site. Despite the absence of this site, the pharmacological properties of the human receptor are similar to those of the guinea pig receptor, suggesting N-linked glycosylation to be nonessential for the basic function of the PAF receptor.[125,126]

## 2. PALMITOYLATION

Palmitoylation has been demonstrated to occur at cysteine residues located in the C-terminal cytoplasmic tail of bovine rhodopsin (Cys322 and Cys323),[127,128] the porcine $\alpha_{2A}$-adrenergic (Cys442)[129] and human $\beta_2$-adrenergic (Cys341)[130] receptors. A proposed consensus sequence for palmitoylation, comprising (Leu-X-Cys-X$_n$-Arg/Lys),[131] is found in many G protein-coupled receptors, with the critical Cys residue occurring approximately 11-16 residues distal to the end of TM7.[17] Palmitoylation may be a common feature of most members of the receptor superfamily. However, the available data have not provided a clear definition of the functional significance of this post-translational modification.

Chemical removal of palmitate from rhodopsin has been reported to enhance coupling to the G protein transducin, $G_T$,[132] while site-directed mutagenesis of bovine opsin, by introduction of the mutations Cys322Ser and/or Cys323Ser, has been reported to have no effect of activation of $G_T$.[133] Conversely, the production of a $\beta_2$-adrenergic receptor carrying the mutations Cys341Ala or Cys341Gly, results in dramatic impairment of functional coupling of the receptor to $G_s$.[130,134] The mutant $\beta_2$-adrenergic receptor carrying the Cys341Gly mutation exhibits elevated levels of phosphorylation in the absence of agonist stimulation and displays neither additional phosphorylation nor further uncoupling in the presence of agonist. The properties of the mutant $\beta_2$-adrenergic receptor resemble those of wild-type receptors which have undergone agonist-induced desensitization (see chapter 2 and chapter 3, below), and identify a role for palmitoylation of the $\beta_2$-adrenergic receptor in regulating accessibility of specific receptor sites to the action of intracellular protein kinases.

Mutation of the $\alpha_{2A}$-adrenergic receptor, by replacement of Cys442 with either Ala or Ser, or mutation of the analogous Cys residue of either the rat m1 muscarinic acetylcholine (Cys435Ser) or porcine m2 muscarinic acetylcholine (Cys457Gly) receptor, causes no detectable change in receptor function.[13,129,135] It has been suggested that palmitoylation may play different roles at different receptor/G protein interfaces, or may mediate a critical function in all receptors which has not yet been elucidated.[129]

## 3. DISULFIDE BOND FORMATION

Biochemical and site-directed mutagenesis studies indicate that Cys residues occurring in the first and second extracellular loops of bovine rhodopsin (Cys110 and Cys187),[136] the hamster $\beta_2$-adrenergic receptor (Cys106, Cys184, Cys190 and Cys191)[11,137] and the rat m1 muscarinic acetylcholine receptor

(Cys98 and Cys178)[13,138,139] are involved in disulfide bond formation and play a crucial role in maintenance of the structural integrity of these receptors. For example, mutants of the m1 muscarinic acetylcholine receptor carrying the mutations Cys98Ser or Cys178Ser exhibit aberrant ligand binding properties and display no functional activation in response to agonist.[13] Similarly, rhodopsin molecules carrying the mutations Cys110Ser and/or Cys187Ser are expressed at reduced levels, are glycosylated abnormally and are unable to bind the ligand 11-*cis*-retinal.[136] In addition, one instance of congenital color blindness, congenital blue cone monochromacy, has been linked to a point mutation of an opsin gene which leads to loss of the conserved Cys residue in the second intradiscal loop (see chapter 4).[140]

The importance of the Cys residues which participate in disulfide bond formation is highlighted by conservation of Cys residues in corresponding locations in the majority of G protein-coupled receptors. However, it is of interest to note that disulfide bond formation between extracellular Cys residues in the $\beta_2$-adrenergic receptor has recently been shown to occur not between the conserved residues Cys106 and Cys184, but to involve disulfide linkage of Cys106 with Cys191 and Cys184 with Cys190 (see chapter 1, Fig. 1.5). High affinity antagonist binding correlates with formation of the disulfide bond between Cys106 and Cys191, while the second disulfide bond, between Cys184 and Cys190, is believed to play a role in agonist affinity.[137]

## G PROTEIN COUPLING

Identification of structural determinants of importance for the coupling of receptors with G proteins has involved the application of a variety of molecular biological approaches. Regions which are essential for coupling have been identified using deletion and site-directed mutagenesis, while peptide competition experiments have given an indication of receptor segments which interact directly with G proteins.

The question of specificity of G protein coupling has been addressed with the construction of chimeric receptors between receptor subtypes that couple to different G proteins.

Extensive analyses with rhodopsin and the $\beta_2$-adrenergic receptor have established the involvement of the second and third intracellular loops and the proximal region of the C-terminal cytoplasmic domain of G protein-coupled receptors in G protein interactions. The experimental data which have contributed to delineation of these regions as interaction sites are discussed in a number of specific review articles.[7,12,14,16,17,141-143] The reader is referred to these articles for a more detailed consideration than is possible within the scope of this review.

Biochemical, mutagenesis and peptide competition studies have identified segments of both the second and third cytoplasmic loops and the C-terminal cytoplasmic domain in the interaction of rhodopsin with $G_T$.[144-148] For activation of $G_T$, there is a specific requirement for an eight amino acid sequence, encompassing amino acid residues 143-150 within the second cytoplasmic loop of bovine rhodopsin, as well as a requirement for residues 244-249 and involvement of residues 236-239 within the third cytoplasmic loop. Residues of particular importance are the conserved Glu134 and Arg135 at the cytoplasmic face of the third TM helix, extending into the second cytoplasmic loop (see chapter 1, Fig. 1.5), and the residues Ser240, Thr243 and Lys248 in the third cytoplasmic loop.[147,149]

Deletion and site-directed mutagenesis experiments have identified similar regions of importance for coupling of the $\beta_2$-adrenergic receptor to $G_s$. There is an absolute requirement for an eight amino acid segment, corresponding to residues 222-229 in the N-terminal membrane-proximal portion of the third intracellular loop, for G protein coupling. In addition, marked impairment of functional coupling to $G_s$ is observed with deletion of 12 amino acids, comprising residues 258-270, from the C-terminal membrane-proximal portion of the third intracellular loop, or deletion of segments of the C-terminal cytoplasmic domain proximal to TM7. Membrane anchorage of the C-terminal cytoplasmic domain of the $\beta_2$-adrenergic receptor, mediated by palmitoylation of Cys341 (see Section 3, above), is also required for optimal coupling of this receptor to $G_s$. The second intracellular loop also plays a role in receptor-G protein interactions,[150,151] and a general requirement for a hydrophobic amino acid residue in the middle of the second intracellular loop, corresponding to Phe139 of the $\beta_2$-adrenergic, Leu131 of the human m1 and Leu174 of the human m3 muscarinic acetylcholine receptors, has recently been identified.[152] As observed for the interaction of rhodopsin with $G_T$, the conserved Asp residue at the cytoplasmic interface of TM3 with the second intracellular loop is important in the coupling of the $\beta_2$-adrenergic receptor to $G_s$.[153,154] The significance of conservation of this residue in the majority of G protein-coupled receptors is highlighted by demonstration of its importance for the coupling of the $\alpha_{2A}$-adrenergic receptor to activation of inwardly rectifying $K^+$ channels and the coupling of m1 muscarinic acetylcholine and angiotensin II $AT_1$ receptors to the stimulation of phosphatidyl inositol (PI) metabolism.[155-158] However, it has also been shown in the rat LH/CG receptor, where the conserved acidic residue is Glu441, that receptor coupling to $G_s$ is unaffected by the mutation Glu441Gln, and that conservation an acidic residue in this location is important for localization of receptors to the cell surface.[159]

Receptor domains of importance for G protein coupling have been identified in a number of receptors, and both similarities and differences from observations made with rhodopsin and the $\beta_2$-adrenergic receptor have been reported. For example, in the $\alpha_{2A}$-adrenergic receptor, it is the membrane-proximal regions of the second intracellular loop and the C-terminal membrane-proximal segment of the third intracellular loop which are implicated in G protein interactions,[160] while site-directed mutagenesis of the TSH receptor has identified a role for the first intracellular loop, as well as C-terminal membrane-proximal segments of the second and third intracellular loops in functional coupling of the TSH receptor.[161,162] Peptide competition and site-directed mutagenesis experiments with the N-formyl peptide receptor indicate lack of a critical role for the third intracellular loop in functional coupling but have implicated the second intracellular loop and the C-terminal cytoplasmic tail of the receptor in G protein interactions.[163-165] Alternatively spliced variants of the pituitary adenylyl cyclase-activating peptide (PACAP) type-I (PACAPR-2) receptor, on the other hand, exhibit differential signaling properties attributable to the insertion of a short cassette of amino acid residues within the third intracellular loop.[166] Isoforms of the somatostatin $SRIF_{1A}$ receptor, which differ in the C-terminal cytoplasmic domain, exhibit differential coupling efficiency to inhibition of adenylyl cyclase activity.[167] Differences in coupling efficiency and G protein selectivity characterize splice variants of the prostaglandin E ($PGE_2$) $EP_3$ receptor, which also differ in the C-terminal cytoplasmic tail of the receptor.[168,169]

Experiments with chimeric receptors, constructed between subtypes of adrenergic and muscarinic acetylcholine receptors which couple to different G proteins and effector systems, have provided evidence for the existence of multiple determinants which are of significance in the selectivity and specificity of coupling interactions. For example, a chimeric m1 muscarinic acetylcholine receptor substituted only in the N-terminal membrane-proximal segment of the third intracellular loop with the corresponding sequence of the $\beta_2$-adrenergic receptor exhibits acquisition of $G_s$ coupling capabilities.[170] This is consistent with the observation that a 15 amino acid peptide comprising this sequence of the $\beta_2$-adrenergic receptor will effect activation of $G_s$ in vitro.[171] However, this sequence does not represent a singular specificity determinant, since $G_s$ coupling is retained in a chimeric

$\beta_2$-adrenergic receptor substituted in the N-terminal membrane-proximal segment of the third intracellular loop with corresponding sequences of $\alpha_{1B}$- or $\alpha_{2A}$-adrenergic receptors.[151,172-174] The C-terminal membrane-proximal segment of the third intracellular loop of the $\beta_2$-adrenergic receptor plays an important role in coupling of this receptor to $G_s$, since its substitution with corresponding sequences of the $\alpha_{1B}$- or $\alpha_{2A}$-adrenergic receptors completely disrupts coupling to $G_s$.[173,174] However, substitution of this sequence, along with substitution of the N-terminal membrane proximal segment of the third intracellular loop and the proximal C-terminal cytoplasmic tail, with corresponding sequences of the $\alpha_{2A}$-adrenergic receptor, results in a chimeric $\beta_2$-adrenergic receptor which retains productive coupling to $G_s$, but has acquired the capability of coupling to $G_i$, which is characteristic of the $\alpha_{2A}$-adrenergic receptor.[173] Such studies highlight the importance of the structural context within which molecular determinants may function to bring about specificity in recognition processes.

Involvement of multiple sequence determinants in G protein recognition has also been established for other receptors. For example, as observed with adrenergic receptor subtypes, membrane-proximal segments of the third intracellular loop have been identified as being of primary importance in G protein selectivity by muscarinic acetylcholine receptor subtypes.[22] Of critical importance is a Tyr residue flanking TM5, which is conserved among the m1, m3 and m5 muscarinic acetylcholine receptors, as well as many other receptors which couple to the stimulation of PI metabolism.[175] In addition, the activation properties of chimeric $\beta_2$-adrenergic/m1 muscarinic acetylcholine receptors have identified a role for the second intracellular loop of the m1 muscarinic acetylcholine receptor in proper recognition of G proteins mediating PI metabolism.[170] Similarly, in metabotropic glutamate receptors (mGluR), which bear minimal sequence homology to the majority of G protein-coupled receptors (see chapter 1, Figures 1.5 and 1.7), specific stimulation of PI metabolism by mGluR1c is dependent on the C-terminal membrane-proximal segment of the second intracellular loop and the membrane-proximal region of the C-terminal cytoplasmic tail.[176]

The membrane-proximal segments of the third intracellular loop of adrenergic, muscarinic acetylcholine, glycoprotein hormone, metabotropic glutamate and other receptors are particularly rich in charged amino acid residues[31,176] and may form cationic amphipathic α-helical extensions of the adjacent TM domains.[25] Such structures exhibit one charged and one hydrophobic face and are characteristic of a naturally occurring peptide, mastorparan, which is a major component of wasp venom and will act as a receptor mimetic in the activation of G proteins.[177] Amphipathic helical structures may also be predicted to occur in the cytoplasmic C-terminus of the receptor for insulin-like growth factor (IGF)-II, which appears to activate a $Ca^{2+}$ channel in a murine fibroblast cell line by interaction with a $G_i$ protein.[178] A synthetic peptide representing one of these domains of the IGF-II receptor, while bearing no primary sequence homology to domains of G protein coupled receptors which are implicated in interaction with G proteins, has been shown to be capable of activation of purified $G_{i-2}$ protein reconstituted into phospholipid vesicles.[179] Similarly, peptides corresponding to the membrane-proximal segments of the third intracellular loop of the $\beta_2$-adrenergic receptor will specifically activate $G_s$, but not $G_i$ or $G_o$,[171] and chimeric IGF-II receptors where a 14 amino acid residue $G_i$-activating domain is replaced by the C-terminal membrane proximal segment of the third cytoplasmic loop of the $\beta_2$-adrenergic receptor exhibit conversion of G protein specificity from $G_i$ to $G_s$.[180] It has also been shown that amphipathic α-helical structure alone does not represent the main structural determinant for receptor-G protein interaction,[181] suggesting that there may be a direct contribution of specific amino acid residues within these regions in determination of receptor-G protein coupling selectivity.

Site-directed mutagenesis studies of the $\beta_2$-adrenergic[182] and m1 muscarinic acetylcholine[183] receptors indicate that the majority of the charged residues contained in these putative helical segments do not play a specific role in receptor-G protein coupling. However, the introduction of mutations which would be predicted to disrupt charge separation on the putative N-terminal membrane-proximal amphipathic helix of the third intracellular loop results in abrogation of coupling of the m3 muscarinic acetylcholine receptor to stimulation of PI metabolism.[184] In addition, specific hydrophobic interactions may play a role in mediating G protein interactions, since replacement of hydrophobic amino acid residues in this region of the $\beta_2$-adrenergic receptor with Leu results in a receptor which is poorly coupled to $G_s$.[182]

Site-directed mutagenesis of the LH/CG receptor in the C-terminal membrane-proximal segment of the third intracellular loop has also provided evidence for lack of requirement for specific charged residues in coupling of this receptor to $G_s$. However, these residues have been shown to be important for optimal expression of the receptor at the cell surface.[185] On the other hand, a pivotal role in functional activation of the TSH receptor has recently been identified for an Ala residue which occurs in this region. Mutant rat TSH receptors carrying the substitutions Ala623Glu or Ala623Lys exhibit a selective loss of coupling to PI metabolism, without loss of coupling to $G_s$.[186] Conversely, a naturally occurring mutation of the human TSH receptor, Ala623Ile, which is one of two mutations identified as underlying the disease hyperfunctioning thyroid adenoma, gives rise to constitutive activation of the TSH receptor and elevation of $G_s$-mediated adenylyl cyclase activity (see chapter 4).[187]

Site-directed mutagenesis of an Ala residue located in a comparable region of the $\alpha_{1B}$-adrenergic receptor, with independent substitution of all possible 19 amino acids for the residue Ala293, also gives rise to receptors with altered signaling properties. In this instance, all mutant receptors exhibit some degree of constitutive activation of the PI metabolism signaling pathway.[188] It has subsequently been shown that mutation of specific residues within this region of both $\alpha_{2A}$- and $\beta_2$-adrenergic receptors also results in constitutive activation.[189,190] Constitutively active $\beta_2$-adrenergic receptors exhibit molecular properties comparable to those of agonist-occupied receptors, including susceptibility to phosphorylation that is normally associated with receptor desensitization (see next Section).[191] These data have identified a more general role for specific sequence determinants within this region of the adrenergic receptors in constraining receptor activity so that functional activation of G proteins does not occur when receptors are not interacting with ligand.[192] An analogous role for the third intracellular loop in negative regulation of receptor activity has recently been identified in the yeast a mating factor receptor.[193]

Constitutively active mutant receptors have also been identified as underlying other pathophysiological conditions. These include one form of autosomal dominant retinitis pigmentosa and familial male precocious puberty, involving the visual pigment rhodopsin and LH/CG receptors, respectively (see chapter 4). Mutations in these receptors do not occur in the C-terminal membrane-proximal segment of the third cytoplasmic loop. However, a closer examination of these and other constitutively active receptors (see above and chapter 4) will undoubtedly increase our understanding of allosteric transitions elicited by ligand binding which represent conversion of receptors from inactive to active states.

# RECEPTOR DESENSITIZATION, SEQUESTRATION AND DOWN-REGULATION

## 1. DESENSITIZATION

Receptor desensitization refers to the phenomenon whereby receptors become refractory to further stimulation after an initial response, despite the continued presence

of a stimulus of constant intensity. A[?] exposure to agonists, receptor-G protein interactions become attenuated due to rapid uncoupling of receptors from their cognate G proteins. As discussed in chapter 2, both agonist-activated and other receptors within the same cell are subject to desensitization in response to exposure to a single agonist. These processes are referred to as homologous and heterologous desensitization, respectively, and phosphorylation of receptors plays a critical role in both processes.

Phosphorylation occurs at Ser and Thr residues predominantly in the third intracellular loop and C-terminal domain of G protein-coupled receptors (see chapter 1, Figs. 1.4 and 1.5). Biochemical and site-directed mutagenesis studies with a number of G protein-coupled receptors have established an involvement of G protein-coupled receptor kinases (GRK), including rhodopsin kinase (GRK1) and the $\beta$-adrenergic receptor kinases ($\beta$ARK)-1 (GRK2) and $\beta$ARK-2 (GRK3),[194-196] as well as the effector kinases protein kinase A (cAMP-dependent protein kinase; PKA) and protein kinase C (PKC), in phosphorylation events mediating receptor desensitization.

It is within the scope of this review to describe only salient features pertaining to structural determinants of receptors which are of significance in desensitization processes. These will be discussed particularly within the context of rhodopsin and the $\beta_2$-adrenergic receptor, since these two types of receptor have been the focus of most intensive investigation. The reader is also referred to our earlier discussion of receptor desensitization in chapter 2 and to a number of excellent reviews for a more detailed treatment of experimental data which have contributed to our current understanding of mechanisms underlying receptor desensitization.[14,16,17,197-200]

Phosphorylation of rhodopsin by rhodopsin kinase (GRK1) occurs on multiple residues within the C-terminal domain of the photoreceptor. Phosphorylation sites on rhodopsin have been mapped with some precision and up to nine phosphate groups may be incorporated.[201] However, only one ıps are believed to be ınt to the inactivation y correspond to phos- 38 and either Ser343 ₃ rhodopsin.[202] Simi- ... ..ıı of the β₂-adrenergic receptor by βARK (GRK2, GRK3) occurs within the C-terminal cytoplasmic domain. Phosphorylation by βARK appears to require the presence of acidic amino acid residues located two or three residues N-terminal to a Ser or Thr residue,[203] and sites of functional significance in the human β₂-adrenergic receptor may be Ser355, Ser356, Thr360, Ser364, Ser401 and/or Ser411 (see chapter 1, Fig. 1.5).[199,204] Phosphorylation by βARK of other receptors, such as the α₂ₐ-adrenergic and m2 muscarinic acetylcholine receptors, appears to occur on residues within the third intracellular loop.[205,206]

Consensus sites for phosphorylation by PKA, represented by the sequence (Arg/Lys-X-(X)-Ser-X),[207] occur in two locations within the β₂-adrenergic receptor. One site corresponds to phosphorylation of Ser262 in the third intracellular loop, and the other site represents phosphorylation of Ser346 within the C-terminal cytoplasmic domain, close to TM7 (see chapter 1, Fig. 1.5). Substitution mutagenesis at these sites, with replacement of Ser residues by Ala or Gly, results in reduced phosphorylation and impaired receptor desensitization.[208] The properties of deletion mutant constructs have allowed identification of the phosphorylation site in the third intracellular loop, involving phosphorylation of Ser262, as being sufficient for PKA-mediated desensitization of the hamster β₂-adrenergic receptor.[209] However, within the β-adrenergic receptor family, the degree of PKA-mediated desensitization may be correlated with the number of PKA phosphorylation sites. Thus, the β₂-adrenergic receptor, which has two sites for PKA-mediated phosphorylation, exhibits pronounced desensitization. The β₁-adrenergic receptor exhibits considerably less desensitization, reflecting the presence of only one PKA phosphorylation site, while the β₃-adrenergic

receptor, which lacks PKA phosphorylation sites, does not undergo PKA-mediated desensitization.[210]

PKA plays a major role in heterologous desensitization of the β₂-adrenergic receptor. Phosphorylation occurs in the presence of low (nanomolar) concentrations of agonist, in response to elevation of intracellular cAMP levels mediated by activation of other receptors which couple through Gₛ to stimulation of adenylyl cyclase activity (see chapter 2). The mechanism by which phosphorylation of the receptor by PKA results in functional uncoupling from Gₛ has not been elucidated. However, PKA phosphorylation sites are located adjacent to cationic regions of the receptor which are believed to be involved in coupling with Gₛ, leading to the suggestion that the introduction of a negatively charged phosphate group may interfere directly with the G protein association.[14] Other G protein-coupled receptors subject to regulation by PKA-mediated phosphorylation include dopamine D₁[211] and PGE₂ EP₁[212] receptors.

Homologous desensitization of the β₂-adrenergic receptor involves the catalytic activities of both βARK and PKA. Phosphorylation mediated by βARK occurs in response to exposure of the receptor to higher (micromolar) concentrations of agonist and is specific for agonist-occupied receptor. Receptors phosphorylated by βARK exhibit enhanced affinity for the inhibitory protein β–arrestin (see chapter 2), whose binding interferes with interaction of the receptors with Gₛ, thereby creating the uncoupled or desensitized state.[213,214] Phosphorylation of receptors by βARK is more rapid than phosphorylation catalyzed by PKA and quantitatively represents the most significant factor contributing to agonist-induced rapid desensitization of the β₂-adrenergic receptor.[200,215]

In addition to rhodopsin and the β₂-adrenergic receptor, a role for phosphorylation mediated by GRKs has been identified for the α₂ₐ-adrenergic, m2 muscarinic acetylcholine and adenosine A₁ receptors, which couple through Gᵢ to inhibition of adenylyl cyclase activity, for the substance P receptor, which couples through $G_{q/11}$ to stimulate PI metabolism and for the thrombin receptor, which couples to a variety of intracellular effector systems.[195,216-218] However, βARK-mediated phosphorylation may not be important in homologous desensitization of other receptors. For example, desensitization of $β_1$-adrenergic receptors in response to agonist stimulation occurs as the result of PKA-mediated events and does not appear to involve βARK.[219] Other receptors subject to homologous desensitization include the dopamine $D_1$, serotonin $5HT_4$, vasopressin $V_2$, parathyroid hormone (PTH), FSH, LH/CG and TSH receptors which couple to $G_s$, the dopamine $D_2$ and serotonin $5HT_{1B}$ receptors which couple to $G_i$, as well as the m3 muscarinic acetylcholine, histamine $H_1$, serotonin $5HT_{2A}$, angiotensin II, cholecystokinin (CCK), complement C5a, N-formyl peptide, NKA, NKB, thromboxane $A_2/PGH_2$ and thrombin receptors which couple to PI metabolism, and the odorant receptors.[200,220-225] The involvement of GRK-mediated phosphorylation in homologous desensitization of these receptors remains to be defined.

Potential phosphorylation sites for PKC, comprising Ser or Thr residues flanked on both sides by basic amino acid residues,[226] occur at Ser261 and Ser262 within the third intracellular loop, and at Ser345 and Ser346 within the C-terminal cytoplasmic domain, of the human $β_2$-adrenergic receptor (see chapter 1, Fig. 1.5). Deletion mutagenesis or substitution of Ala for Ser at these locations abrogates the functional uncoupling from $G_s$ that is observed in wild-type receptors in response to activation of PKC by phorbol ester treatment. These data suggest the involvement of PKC-mediated phosphorylation also in heterologous desensitization of the wild-type $β_2$-adrenergic receptor.[227,228] There is a correspondence of sites for PKC-mediated phosphorylation with sites for PKA-mediated phosphorylation and, in the $β_2$-adrenergic receptor, a corresponding preference for phosphorylation by PKC at the site within the third intracellular loop.[200,227] This suggests

a similar mechanism for direct interference with G protein coupling by PKC-mediated phosphorylation of the receptor. Other receptors subject to PKC-mediated desensitization, in addition to homologous desensitization, include rhodopsin, histamine $H_1$, serotonin $5HT_{2A}$, angiotensin II, complement C5a and thrombin receptors.[200,229,230]

## 2. SEQUESTRATION

In response to agonist stimulation, usually within the time frame of several minutes, receptors which have become desensitized are reversibly sequestered. The functional significance of receptor sequestration is not clear and the process itself is poorly understood.

In the case of $β_2$-adrenergic receptors, sequestered receptors remain detectable by hydrophobic ligands but become increasingly inaccessible to hydrophilic ligands.[16] There is biochemical, pharmacological and immunocytochemical evidence that agonist-induced sequestration involves internalization of receptors.[16,200] Sequestered $β_2$-adrenergic receptors have been shown to be colocalized in intracellular vesicles with transferrin receptors,[231] indicating that the same endosomal sorting pathway may be used in sequestration processes as is used for receptors such as transferrin receptors, which are constitutively recycled. Consistent with this, the use of strains of yeast expressing a temperature-sensitive clathrin heavy chain has allowed demonstration of an important role for clathrin in internalization of yeast pheromone receptors.[232]

Sequestration may represent a mechanism for resensitization of receptors.[199,200] This is consistent with the observations that $β_2$-adrenergic receptors sequestered in response to exposure to agonist are less phosphorylated than those remaining in the plasma membrane,[233] and that inhibition of $β_2$-adrenergic receptor sequestration, while having no detrimental effect on receptor desensitization,[234] does prevent resensitization of receptors after removal of agonist.[235]

While it is known that receptor sequestration is initiated by agonist occupancy of receptors, the structural determinants which play a role in receptor sequestration have not been fully defined. Site-directed mutagenesis of the human m1 muscarinic acetylcholine receptor has established that agonist-induced receptor sequestration requires receptor domains which are similar to and overlap with domains of importance in G protein coupling.[236,237] However, the properties of substitution or deletion mutant constructs of a number of receptors, including $\beta_2$-adrenergic, m1 muscarinic acetylcholine and bombesin/gastrin-releasing peptide (GRP) receptors, indicate that sequestration does not appear to require coupling to G proteins, nor is it dependent on phosphorylation associated with receptor desensitization.[174,208,237-241]

Agonist-induced sequestration is not impaired in mutant $\beta_2$-adrenergic receptors where all 11 Ser or Thr residues which are potential sites for $\beta$ARK-mediated phosphorylation within the C-terminal cytoplasmic domain have been substituted with Gly or Ala residues.[242] However, site-directed mutagenesis of one Thr and three Ser residues, which occur within a stretch of ten amino acids encompassed by residues Ser355 to Ser364 within the C-terminal cytoplasmic domain of this receptor, does affect rapid regulation of receptor function. This includes receptor sequestration, as well as agonist-induced phosphorylation and desensitization (see Section 1, above), and may involve conformational constraints imposed on the receptor by this sequence.[204] Agonist-induced internalization of the murine thyrotropin-releasing hormone (TRH) and human m1, m2 and m3 muscarinic acetylcholine receptors is dependent on similar short stretches of amino acids which are rich in Ser and Thr residues. The regions of importance are located in the C-terminal cytoplasmic tail of the TRH receptor and in the middle of the third intracellular loop of the muscarinic acetylcholine receptors.[243,244] The involvement of Ser and Thr residues in mechanisms underlying agonist-induced internalization of receptors is not clear, but it is of interest to note that specific Ser and Thr residues in the C-terminal cytoplasmic do-

main have been shown to be essential for internalization of the bombesin/GRP BB$_2$ receptor.[241] Clearly, the possibility that phosphorylation, catalyzed by an as yet unidentified kinase(s), mediates regulation of at least some pathway(s) of receptor internalization[244] cannot be excluded. Of additional critical importance for agonist-mediated internalization of the TRH receptor are Cys residues within the C-terminal cytoplasmic domain, which may be sites for receptor palmitoylation[243] This may reflect a more general role for conformational determinants in receptor sequestration mechanisms. In addition, a highly conserved Tyr residue, which occurs as part of a conserved sequence motif in TM7 of the majority of G protein-coupled receptors (see chapter 1, Fig. 1.5), has recently been shown to be essential for agonist-mediated receptor sequestration, while not playing a role in coupling or down-regulation processes.[245]

## 3. DOWN-REGULATION

In response to agonist stimulation in the longer term, receptor levels become down-regulated. There is an irreversible loss of receptors from the plasma membrane due to both the internalization and degradation of receptors, and reduction in receptor mRNA levels. Down-regulation of receptors occurs more slowly than sequestration. Receptor loss is most rapid during the first four hours of exposure to agonist and approaches steady state by 24 hours of continuous exposure to agonist.[246]

As described for desensitization processes, receptor down-regulation may be homologous, resulting in removal of receptors specific for the stimulating agonist, or heterologous, as observed for the $\beta_2$-adrenergic receptor in response to the application of cAMP analogs or elevation of intracellular cAMP levels through the use of pharmacological agents.[200]

Agonist-induced down-regulation of $\beta_2$-adrenergic receptors is dependent on G protein coupling and is impaired if coupling is defective due to mutations in either the receptor or G$_s$.[174,247,248] The mechanisms involved have not been defined, but could include the generation of a G protein-mediated signal or the preferential degradation of receptor-G protein complexes[200] There is also evidence that PKA-mediated receptor phosphorylation may play a role in down-regulation of $\beta_2$-adrenergic receptors, since mutant receptors lacking PKA phosphorylation sites exhibit slower down-regulation than wild-type receptors.[246] Similarly, site-specific mutagenesis of Thr residues, including potential PKC phosphorylation sites, in the C-terminal cytoplasmic domain of the m3 muscarinic acetylcholine receptor leads to a significant decrease in agonist-induced down-regulation.[249] However, studies with mutant $\beta_2$-adrenergic receptors lacking $\beta$ARK phosphorylation sites, indicate that agonist-induced $\beta$ARK-mediated phosphorylation does not play a role in down-regulation processes.[174,204]

The relationship between receptor sequestration and down-regulation mechanisms remains ill-defined. Site-directed mutagenesis has provided evidence for an involvement of Tyr residues located within the C-terminal cytoplasmic domain of the $\beta_2$-adrenergic receptor in down-regulation processes, with a specific requirement for either Tyr350 or Tyr354 for maintenance of normal agonist-induced down-regulation (see chapter 1, Fig. 1.5).[250,251] While similar Tyr residues have been shown to be important for the internalization of other receptors, such as the low density lipoprotein (LDL) receptor, which are recycled to the plasma membrane, $\beta_2$-adrenergic receptors carrying the mutations Tyr350Ala and Tyr354Ala are not impaired in receptor sequestration processes.[250] Mutant receptors which are impaired in G protein coupling and show blunted down-regulation are also sequestered normally.[174] Conversely, receptors which are impaired in sequestration processes due to mutation of Ser and Thr residues within a stretch of 10 amino acids in the C-terminal cytoplasmic domain of the $\beta_2$-adrenergic receptor (see Section 2, above) are down-regulated normally.[204]

Immunocytochemical studies addressing the requirement for internalization of receptors in longer-term down-regulation and degradation processes have yielded conflicting results.[252,253] However, in mechanistic terms, it would appear plausible that receptor internalization, which characterizes the sequestration process, would precede down-regulation attributable to degradation. In order to reconcile the apparent independence of internalization and degradative processes, it has been suggested that in normal circumstances, the majority of sequestered receptors may be recycled to the plasma membrane and only a small proportion of sequestered receptors are directed to lysosomal compartments for degradation. Thus, normal down-regulation would reflect degradation of only a small component of internalized receptors and may still occur in situations where sequestration is impaired.[231]

A progressive reduction in steady state levels of receptor mRNA in response to exposure to agonist has been reported for a number of receptors, including the muscarinic acetylcholine,[254-256] TRH,[257] LH/CG[52] and TSH[258] receptors, while it has been established for the $\beta_2$-adrenergic receptor that both homologous and heterologous down-regulation processes are associated with reduction of receptor mRNA.[246,259] For the $\beta_2$-adrenergic,[260] LH/CG[261] and TRH[262] receptors, it has been shown that this is due to mRNA destabilization, which may involve the binding of specific proteins[263] or be dependent upon specific mRNA sequence determinants.[264] Recovery of the full complement of receptors by cells which have undergone receptor down-regulation may take on the order of days or weeks and is dependent on new protein synthesis.[265,266]

## SUMMARY

Ligand binding and receptor activation determinants have now been described for a number of receptors and receptor subtypes. The metabotropic glutamate receptors, like the adrenergic and muscarinic acetylcholine receptors, bind a small ligand. However, unlike the adrenergic and muscarinic acetylcholine receptors where ligand binding occurs within the mostly hydrophobic TM region, metabotropic glutamate receptors exhibit ligand recognition properties within the large N-terminal extracellular domain. Receptors for the multimeric glycoprotein hormones FSH, LH/CG and TSH are also characterized by a large N-terminal extracellular domain, within which ligand binding determinants have been identified.

There are a large number of peptides of diverse structure which interact with G protein-coupled receptors (see chapter 1, Table 1.1). The definition of ligand recognition determinants for several peptide ligands is giving an indication of an involvement of both extracellular and TM domains in recognition properties, with particular features characterizing discrete ligand-receptor interactions. For example, multiple domains contribute to ligand recognition and discrimination in receptors for the tachykinin peptides substance P, NKA (substance K) and NKB (neuromedin K). Also of interest is the receptor for the peptide thrombin, which exhibits a novel and unique activation mechanism. The interaction of thrombin with its receptor results in cleavage of the receptor at its existing N-terminus, generating a new N-terminus which functions as a tethered ligand to effect receptor activation.

The contributions to receptor function of post-translational modifications of receptor sequence, including N-linked glycosylation, palmitoylation and intramolecular disulfide bond formation, have now been examined in a number of receptors, with identification of a role in ligand binding and receptor activation only in certain cases.

Attempts to define G protein coupling determinants encompass studies directed towards delineation of recognition features for particular G proteins. In addition, a number of mutant receptors have been constructed or identified which exhibit constitutive activation. Assessment of the functional capabilities of such receptors has shed new light on mechanisms for regulation of receptor function and is of relevance

to molecular events underlying not only ligand binding and receptor activation, but also G protein interaction and stimulation.

Finally, the efficacy of signal transduction relies crucially on desensitization mechanisms, manifested as both homologous and heterologous receptor desensitization. A number of structural determinants have been identified which mediate this important regulatory process.

## REFERENCES

1. Oprian DD. The ligand-binding domain of rhodopsin and other G protein-linked receptors. J Bioenerg Biomemb 1992; 24:211-217.

2. Hargrave PA, McDowell JH, Siemiatkowski JEC et al. The carboxy-terminal one-third of bovine rhodopsin: its structure and function. Vision Res 1982; 22:1429-1438.

3. Thomas DD, Stryer L. Transverse location of the retinal chromophore of rhodopsin in rod outer segment disc membranes. J Mol Biol 1982; 154:145-157.

4. Hargrave PA, McDowell JH. Rhodopsin and phototransduction: a model system for G protein-linked receptors. FASEB J 1992; 6:2323-2331.

5. Khorana HG. Rhodopsin, photoreceptor of the rod cell. An emerging pattern for structure and function. J Biol Chem 1992; 267:1-4.

6. Hargrave PA, Hamm HE, Hofmann KP. Interaction of rhodopsin with the G-protein, transducin. BioEssays 1993; 15:43-50.

7. Hamm HE. Molecular interactions between the photoreceptor G protein and rhodopsin. Cell Mol Neurobiol 1991; 11:563-578.

8. Farahbakhsh ZT, Hideg K, Hubbell WL. Photoactivated conformational changes in rhodopsin: a time-resolved spin label study. Science 1993; 262:1416-1419.

9. Weitz CJ, Nathans J. Histidine residues regulate the transition of photoexcited rhodopsin to its active conformation, metarhodopsin II. Neuron 1992; 8:465-472.

10. Dixon RAF, Sigal IS, Rands E et al. Ligand binding to the β-adrenergic receptor involves its rhodopsin-like core. Nature 1987; 326:73-77.

11. Dixon RAF, Sigal IS, Candelore MR et al.

Structural features required for ligand binding to the β-adrenergic receptor. EMBO J 1987; 6:3269-3275.

12. Ostrowski J, Kjelsberg MA, Caron MG et al. Mutagenesis of the β₂-adrenergic receptor: how structure elucidates function. Annu Rev Pharmacol Toxicol 1992; 32:167-183.

13. Saravese TM, Wang C-D, Fraser CM. Site-directed mutagenesis of the rat m₁ muscarinic acetylcholine receptor. Role of conserved cysteines in receptor function. J Biol Chem 1992; 267:11439-11448.

14. Dohlman HG, Thorner J, Caron MG et al. Model systems for the study of seven-transmembrane-segment receptors. Annu Rev Biochem 1991; 60:653-688.

15. Tota MR, Candelore MR, Dixon RAF et al. Biophysical and genetic analysis of the ligand-binding site of the β-adrenoceptor. Trends Pharmacol Sci 1991; 12:4-6.

16. Kobilka B. Adrenergic receptors as models for G protein-coupled receptors. Annu Rev Neurosci 1992; 15:87-114.

17. Saravese TM, Fraser CM. *In vitro* mutagenesis and the search for structure-function relationships among G protein-coupled receptors. Biochem J 1992; 283:1-19.

18. Strosberg AD. Structure, function, and regulation of adrenergic receptors. Prot Sci 1993; 2:1198-1209.

19. Venter JC, Fraser CM. Structure and molecular biology of transmitter receptors. Am Rev Respir Dis 1990; 141:S99-S105.

20. Brann MR, Klimkowski VJ, Ellis J. Structure/function relationships of muscarinic acetylcholine receptors. Life Sci 1993; 52:405-412.

21. Wess J. Molecular basis of muscarinic acetylcholine receptor function. Trends Pharmacol Sci 1993; 14:308-313.

22. Wess J. Mutational analysis of muscarinic acetylcholine receptors: structural basis of ligand/receptor/G protein interactions. Life Sci 1993; 53:1447-1463.

23. Strader CD, Sigal IS, Dixon RAF. Mapping the functional domains of the β-adrenergic receptor. Am J Respir Cell Mol Biol 1989; 1:81-86.

24. Green SA, Cole G, Jacinto M et al. A polymorphism of the human β₂-adrenergic receptor within the fourth transmembrane domain alters ligand binding and functional properties of the receptor. J Biol Chem 1993; 268:23116-23121.

25. Strader CD, Sigal IS, Dixon RAF. Structural basis of β-adrenergic receptor function. FASEB J 1989; 3:1825-1832.

26. Frielle T, Daniel KW, Caron MG et al. Structural basis of β-adrenergic receptor subtype specificity studied with chimeric β₁/β₂-adrenergic receptors. Proc Natl Acad Sci USA 1988; 85:9494-9498.

27. Kobilka BK, Kobilka TS, Daniel K et al. Chimeric α₂-, β₂-adrenergic receptors: delineation of domains involved in effector coupling and ligand binding specificity. Science 1988; 240:1310-1316.

28. Suryanarayana S, Daunt DA, Von Zastrow M et al. A point mutation in the seventh hydrophobic domain of the α₂ adrenergic receptor increases its affinity for a family of β receptor antagonists. J Biol Chem 1991; 266:15488-15492.

29. Wilson AL, Guyer CA, Cragoe EJ et al. The hydrophobic tryptic core of the porcine α₂-adrenergic receptor retains allosteric modulation of binding by Na⁺, H⁺, and 5-amino-substituted amiloride analogs. J Biol Chem 1990; 265:17318-17322.

30. Strosberg AD, Camoin L, Blin N et al. In receptors coupled to GTP-binding proteins, ligand binding and G-protein activation is a multistep dynamic process. Drug Des Discov 1993; 9:199-211.

31. Seeman P. Receptor amino acid sequences of G-linked receptors. First Edition. Toronto: University of Toronto, 1992.

32. Wess J, Gdula D, Brann MR. Site-directed mutagenesis of the m3 muscarinic receptor: identification of a series of threonine and tyrosine residues involved in agonist but not antagonist binding. EMBO J 1991; 10:3729-3734.

33. Wess J, Nanavati S, Vogel Z et al. Functional role of proline and tryptophan residues highly conserved among G protein-coupled receptors studied by mutational analysis of the m3 muscarinic receptor. EMBO J 1993; 12:331-338.

34. Wess J, Bonner TI, Dörje F et al. Delineation of muscarinic receptor domains conferring selectivity of coupling to guanine nucle-

otide proteins and second messengers. Mol Pharmacol 1990; 38:517-523.

35. Lai J, Nunan L, Waite SL et al. Chimeric M1/M2 muscarinic receptors: correlation of ligand selectivity and functional coupling with structural modifications. J Pharmacol Exp Ther 1992; 262:173-180.

36. Wess J, Bonner TI, Brann MR. Chimeric m2/m3 muscarinic receptors: role of carboxy terminal receptor domains in selectivity of ligand binding and coupling to phosphoinositide hydrolysis. Mol Pharmacol 1990; 38:872-877.

37. Wess J, Gdula D, Brann MR. Structural basis of the subtype selectivity of muscarinic antagonists: a study with chimeric m2/m5 muscarinic receptors. Mol Pharmacol 1992; 41:369-374.

38. Drübbisch V, Lameh J, Philip M et al. Mapping the ligand binding pocket of the human muscarinic cholinergic receptor Hm1: contribution of tyrosine-82. Pharm Res 1992; 9:1644-1647.

39. Wess J, Maggio R, Palmer JR. Role of conserved threonine and tyrosine residues in acetylcholine binding and muscarinic receptor activation. A study with m3 muscarinic receptor point mutants. J Biol Chem 1992; 267:19313-19319.

40. Tanabe Y, Masu M, Ishii T et al. A family of metabotropic glutamate receptors. Neuron 1992; 8:169-179.

41. Pin J-P, Waeber C, Prezeau L et al. Alternative splicing generates metabotropic glutamate receptors inducing different patterns of calcium release in Xenopus oocytes. Proc Natl Acad Sci USA 1992; 89:10331-10335.

42. Abe T, Sugihar H, Nawa H et al. Molecular characterization of a novel metabotropic glutamate receptor mGluR5 coupled to inositol phosphate/$Ca^{2+}$ signal transduction. J Biol Chem 1992; 267:13361-13368.

43. Nakajima Y, Iwakabe H, Akazawa C et al. Molecular characterization of a novel retinal metabotropic glutamate receptor mGluR6 with a high agonist selectivity for L-2-amino-4-phosphonobutyrate. J Biol Chem 1993; 268:11868-11873.

44. Okamoto N, Hori S, Akazawa C et al. Molecular characterization of a new meta-botropic glutamate receptor mGluR7 coupled to inhibitory cyclic AMP signal transduction. J Biol Chem 1994; 269:1231-1236.

45. Takahashi K, Tsuchida K, Tanabe Y et al. Role of the large extracellular domain of metabotropic glutamate receptors in agonist selectivity determination. J Biol Chem 1993; 268:19341-19345.

46. Moyle WR, Bernard MP, Myers RV et al. Leutropin/β-adrenergic receptor chimeras bind choriogonadotropin and adrenergic ligands but are not expressed at the cell surface. J Biol Chem 1991; 266:10807-10812.

47. Tsai-Morris CH, Buczko E, Wand W et al. Intronic nature of the rat luteinizing hormone receptor gene defines a soluble receptor subspecies with hormone binding activity. J Biol Chem 1990; 265:19385-19388.

48. Xie Y-B, Wang HY, Segaloff DL. Extracellular domain of lutropin/choriogonadotropin receptor expressed in transfected cells binds choriogonadotropin with high affinity. J Biol Chem 1990; 265:21411-21414.

49. Reichert LE, Dattatreyamurty B, Grasso P et al. Structure-function relationships of the glycoprotein hormones and their receptors. Trends Pharmacol Sci 1991; 12:199-203.

50. Braun T, Schofield PR, Sprengel R. Amino-terminal leucine-rich repeats in gonadotropin receptors determine hormone selectivity. EMBO J 1991; 10:1885-1890.

51. Dias JA. Recent progress in structure-function and molecular analyses of the pituitary/placental glycoprotein hormone receptors. Biochim Biophys Acta 1992; 1135:278-294.

52. Segaloff DL, Ascoli M. The lutropin/choriogonadotropin receptor... 4 years later. Endocr Rev 1993; 14:324-347.

53. Dattatreyamurty B, Reichert LE. A synthetic peptide corresponding to amino acids 9-30 of the extracellular domain of the follitropin (FSH) receptor specifically binds FSH. Mol Cell Endocrinol 1992; 87:9-17.

54. Nagayama Y, Russo D, Chazenbalk GD et al. Extracellular domain chimeras of the TSH and LH/CG receptors reveal the midregion (amino acids 171-260) to play a vital role in high affinity TSH binding. Biochem Biophys Res Commun 1990;

173:1150-1156.

55. Wadsworth HL, Chazenbalk GD, Nagayama Y et al. An insertion in the human thyrotropin receptor critical for high affinity hormone binding. Science 1990; 249: 1423-1425.

56. Atassi MZ, Manshouri T, Sakata S. Localization and synthesis of the hormone-binding regions of the human thyrotropin receptor. Proc Natl Acad Sci USA 1991; 88:3613-3617.

57. Ohmori M, Endo T, Ikeda M et al. Role of N-terminal region of the thyrotropin (TSH) receptor in signal transduction for TSH or thyroid-stimulating antibody. Biochem Biophys Res Commun 1991; 178:733-738.

58. Morris JC, Bergert ER, McCormick DJ. Structure-function studies of the human thyrotropin receptor. Inhibition of binding of labeled thyrotropin (TSH) by synthetic human TSH receptor peptides. J Biol Chem 1993; 268:10900-10905.

59. Nagayama Y, Wadsworth HL, Chazenbalk GD et al. Thyrotropin-luteinizing hormone/chorionic gonadotropin receptor extracellular domain chimeras as probes for thyrotropin receptor function. Proc Natl Acad Sci USA 1991; 88:902-905.

60. Nagayama Y, Russo D, Chazenbalk GD et al. Seven amino acids (lys-201-lys-211) and 9 amino acids (gly-222 to leu-230) in the human thyrotropin receptor are involved in ligand binding. J Biol Chem 1991; 266:14926-14930.

61. Kosugi S, Ban T, Akamizu T et al. Site-directed mutagenesis of a portion of the extracellular domain of the rat thyrotropin receptor important in autoimmune thyroid disease and nonhomologous with gonadotropin receptors. Relationship of functional and immunogenic domains. J Biol Chem 1991; 266:19413-19418.

62. Kosugi S, Ban T, Akamizu T et al. Further characterization of a high affinity thyrotropin binding site on the rat thyrotropin receptor from idiopathic myxedema patients but not thryoid stimulating antibodies from Graves' patients. Biochem Biophys Res Commun 1991; 180:1118-1124.

63. Roche PC, Ryan RJ, McCormick DJ. Identification of hormone-binding regions of the luteinizing hormone/human chorionic gonadotropin receptor using synthetic peptides. Endocrinology 1992; 131:268-274.

64. Dattatreyamurty B, Reichert LE. Identification of regions of the follitropin (FSH) β-subunit that interact with the N-terminus region (residues 9-30) of the FSH receptor. Mol Cell Endocrinol 1993; 93:39-46.

65. Moyle WR, Campbell RK, Myers RV et al. Co-evolution of ligand-receptor pairs. Nature 1994; 368:251-255.

66. Ji I, Ji TH. Human choriogonadotropin binds to a lutropin receptor with essentially no N-terminal extension and stimulates cAMP synthesis. J Biol Chem 1991; 266:13076-13079.

67. Ji I, Ji TL. Asp[383] in the second transmembrane domain of the lutropin receptor is important for high affinity hormone binding and cAMP production. J Biol Chem 1991; 266:14953-14957.

68. Quintana J, Wang H, Ascoli M. The regulation of the binding affinity of the luteinizing hormone/choriogonadotropin receptor is mediated by a highly conserved aspartate located in the second transmembrane domain of G protein-coupled receptors. Endocrinology 1993; 7:767-775.

69. Horstman DA, Brandon S, Wilson AL et al. An aspartate conserved among G-protein receptors confers allosteric regulation of $\alpha_2$-adrenergic receptors. J Biol Chem 1990; 21590-21595.

70. Ji I, Ji TH. Receptor activation is distinct from hormone binding in intact lutropin-choriogonadotropin receptors and Asp[397] is important for receptor activation. J Biol Chem 1993; 268:20851-20854.

71. Ji I, Zeng H, Ji TH. Receptor activation of and signal generation by the lutropin/choriogonadotropin receptor. Cooperation of Asp[397] of the receptor and αLys[91] of the hormone. J Biol Chem 1993; 268: 22971-22974.

72. Yamano Y, Ohyama K, Chaki S et al. Identification of amino acid residues of rat angiotensin II receptor for ligand binding by site-directed mutagenesis. Biochem Biophys Res Commun 1992; 187:1426-1431.

73. Kong H, Raynor K, Yasuda K et al. A single residue, aspartic acid 95, in the δ opioid

receptor specifies selective high affinity agonist binding. J Biol Chem 1993; 268:23055-23058.

74. Radel SJ, Genco RJ, De Nardin E. Localization of ligand binding regions of the human formyl peptide receptor. Biochem Int 1991; 25:745-753.

75. Perez HD, Holmes R, Vilanders LR et al. Formyl peptide receptor chimeras define domains involved in ligand binding. J Biol Chem 1993; 268:2292-2295.

76. Quehenberger O, Prossnitz ER, Cavanagh SL et al. Multiple domains of the N-formyl peptide receptor are required for high-affinity ligand binding. J Biol Chem 1993; 268:18167-18175.

77. Fathi Z, Benya RV, Shapira H et al. The fifth transmembrane segment of the neuromedin B receptor is critical for high affinity neuromedin B binding. J Biol Chem 1993; 268:14622-14626.

78. Zhu G, Wu LH, Mauzy C et al. Replacement of lysine-181 by aspartic acid in the third transmembrane region of endothelin type B receptor reduces its affinity to endothelin peptides and sarafotoxin 6c without affecting G protein coupling. J Cell Biochem 1992; 50:159-164.

79. Adachi M, Yang Y-Y, Trzeciak A et al. Identification of a domain of $ET_A$ receptor required for ligand binding. FEBS Lett 1992; 311:179-183.

80. Sakamoto A, Yanagisawa M, Sawamura T et al. Distinct subdomains of human endothelin receptors determine their selectivity to endothelin$_A$-selective antagonist and endothelin$_B$-selective agonists. J Biol Chem 1993; 268:8547-8553.

81. Maggio JE. Tachykinins. Annu Rev Neurosci 1988; 11:13-28.

82. Huang R-RC, Yu H, Strader CD et al. Interaction of substance P with the second and seventh transmembrane domains of the neurokinin-1 receptor. Biochemistry 1994; 33:3007-3013.

83. Fong TM, Huang R-RC, Strader CD. Localization of agonist and antagonist binding domains of the human neurokinin-1 receptor. J Biol Chem 1992; 267: 25664-25667.

84. Fong TM, Yu H, Huang R-RC et al. The

extracellular domain of the nuerokinin-1 receptor is required for high-affinity binding of peptides. Biochemistry 1992; 31:11806-11811.

85. Fong TM, Yu H, Strader CD. The extracellular domain of substance P (NK1) receptor comprises part of the ligand binding site. Biophys J 1992; 62:59-60.

86. Yokota Y, Akazawa C, Ohkubo H et al. Delineation of structural domains involved in the subtype specificity of tachykinin receptors through chimeric formation of substance P/substance K receptors. EMBO J 1992; 11:3585-3591.

87. Gether U, Johanson TE, Snider RM et al. Binding epitopes for peptide and nonpeptide ligands on the $NK_1$ (substance P) receptor. Regul Pept 1993; 46:49-58.

88. Gerard NP, Bao L, Xiao-Ping H et al. Molecular aspects of the tachykinin receptors. Regul Pept 1993; 43:21-35.

89. Gether U, Johansen TE, Snider RM et al. Different binding epitopes on the $NK_1$ receptor for substance P and a nonpeptide antagonist. Nature 1993; 362:345-348.

90. Fong TM, Yu H, Ctrader CD. Molecular basis for the species selectivity of the neurokinin-1 receptor antagonists CP-96,345 and RP67580. J Biol Chem 1992; 267:25668-25671.

91. Sachais BS, Snider RM, Lowe JA et al. Molecular basis for the species specificity of the substance P antagonist CP-96,345. J Biol Chem 1993; 268:2319-2323.

92. Fong TM, Cascierei MA, Yu H et al. Amino-aromatic interaction between histidine 197 of the neurokinin-1 receptor and CP 96345. Nature 1993; 362:350-353.

93. Vu T-K, Wheaton VI, Hung DT et al. Domains specifying thrombin-receptor interaction. Nature 1991; 353:674-677.

94. Coughlin SR. Thrombin receptor structure and function. Thromb Haemost 1993; 70:184-187.

95. Vu T-K, Hung D, Wheaton VI et al. Molecular cloning of a functional thrombin receptor reveals a novel proteolytic mechanism of receptor activation. Cell 1991; 64:1057-1068.

96. Hung DT, Wong YH, Vu T-KH et al. The cloned platelet thrombin receptor couples

to at least two distinct effectors to stimulate phosphoinositide hydrolysis and inhibit adenylyl cyclase. J Biol Chem 1992; 267:20831-20834.

97. Ishii K, Hein L, Kobilka B et al. Kinetics of thrombin receptor cleavage on intact cells: relation to signaling. J Biol Chem 1993; 268:9780-9786.

98. Vouret-Craviari V, Obberghen-Schilling V, Rasmussen U et al. Synthetic α-thrombin receptor peptides activate G-protein coupled signalling pathways but are unable to induce mitogenesis. Mol Cell Biol 1992; 3:95-102.

99. Vassallo R, Kieber-Emmons T, Cichowski K et al. Structure-function relationships in the activation of platelet thrombin receptors by receptor-derived peptides. J Biol Chem 1992; 267:6081-6085.

100. Scarborough RM, Naughton M, Teng W et al. Tethered ligand agonist peptides. Structural requirements for thrombin receptor activation reveal mechanism of proteolytic unmasking of agonist function. J Biol Chem 1992; 267:13146-13149.

101. Coller BS, Ward P, Ceruso M et al. Thrombin receptor activating peptides: importance of the N-terminal serine and its ionization state as judged by pH dependence, NMR spectroscopy, and cleavage by aminopeptidase M. Biochemistry 1992; 31: 11713-11720.

102. La Rosa GJ, Thomas KM, Kaufmann ME et al. Amino terminus of the interleukin-8 receptor is a major determinant of receptor subtype specificity. J Biol Chem 1992; 267:25402-25406.

103. Gayle RB, Sleath PR, Srinivason S et al. Importance of the amino terminus of the interleukin-8 receptor in ligand interactions. J Biol Chem 1993; 268:7283-7289.

104. Hébert CA, Chuntharapai A, Smith M et al. Partial functional mapping of the human interleukin-8 type A receptor. Identification of a major ligand binding determinant. J Biol Chem 1993; 268:18549-18553.

105. Buck L, Axel R. A novel multigene family may encode odorant receptors: a molecular basis for odor recognition. Cell 1991; 65:175-187.

106. Levy NS, Bakalyar HA, Reed RR. Signal transduction in olfactory neurons. J Steroid Biochem 1991; 39:633-637.

107. Parmentier M, Libert F, Schurmans S et al. Expression of members of the putative olfactory receptor gene family in mammalian germ cells. Nature 1992; 355:453-455.

108. Selbie LA, Townsend-Nicholson A, Iismaa TP et al. Novel G protein-coupled receptors: a gene family of putative human olfactory receptor sequences. Mol Brain Res 1992; 13:159-163.

109. Ngai J, Dowling MM, Buck L et al. The family of genes encoding odorant receptors in the channel catfish. Cell 1993; 72: 657-666.

110. Kornfeld R, Kornfeld S. Assembly of asparagine-linked oligosaccharides. Annu Rev Biochem 1985; 54:631-664.

111. Rands E, Candelore MR, Cheung AH et al. Mutational analysis of β-adrenergic receptor glycosylation. J Biol Chem 1990; 265:10759-10764.

112. van Koppen CJ, Nathanson NM. Site-directed mutagenesis of the m2 muscarinic acetylcholine receptor. Analysis of the role of *N*-glycosylation in receptor expression and function. J Biol Chem 1990; 265: 20887-20892.

113. Klotz K-N, Lohse MJ. The glycoprotein nature of $A_1$ adenosine receptors. Biochem Biophys Res Commun 1986; 140:406-413.

114. Benovic JL, Staniszewski C, Cerione RA et al. The mammalian β-adrenergic receptor: structural and functional characterization of the carbohydrate moiety. J Recept Res 1987; 7:257-281.

115. Hootman SR, Verme TB, Habara Y. Effects of cycloheximide and tunicamycin on cell surface expression of pancreatic muscarinic acetylcholine receptors. FEBS Lett 1990; 274:35-38.

116. Liu X, Davis D, Segaloff DL. Disruption of potential sites for *N*-linked glycosylation does not impair hormone binding to the lutropin/choriogonadotropin receptor if Asn-173 is left intact. J Biol Chem 1993; 268:1513-1516.

117. Russo D, Chazenbalk GD, Nagayama Y et al. Site-directed mutagenesis of the human thyrotropin receptor: role of asparagine-linked oligosaccharides in the expression of

a functional receptor. Mol Endocrinol 1991; 5:29-33.

118. Remes JJ, Petäjä UE, Tuukkanen KJ et al. Significance of the extracellular domain and the carbohydrates of the human neutrophil N-formyl peptide chemotactic receptor for the signal transduction by the receptor. Exp Cell Res 1993; 209:26-32.

119. Haendler B, Hechler U, Becker A et al. Expression of human endothelin receptor ET$_B$ by *Escherichia coli* transformants. Biochem Biophys Res Commun 1993; 191:633-638.

120. Herzog H, Shine J. Human NPY Y1 receptor expressed in *E. coli* retains its pharmacological properties. DNA and Cell Biology 1994; *in press*.

121. Grisshammer R, Duckworth R, Henderson R. Expression of a rat neurotensin receptor in *Escherichia coli*. Biochem J 1993; 295:571-576.

122. Santer R, Leung YK, Alliet P et al. The role of carbohydrate moieties of cholecystokinin receptors in cholecystokinin octapeptide binding: alteration of binding data by specific lectins. Biochim Biophys Acta 1990; 1051:78-83.

123. Rens-Domiano S, Reisine T. Structural analysis and functional role of the carbohydrate component of somatostatin receptors. J Biol Chem 1991; 266:20094-20102.

124. Chochola J, Fabre C, Bellan C et al. Structural and functional analysis of the human vasoactive intestinal peptide receptor glycosylation. J Biol Chem 1993; 268: 2312-2318.

125. Honda Z, Nakamura M, Miki I et al. Cloning by functional expression of platelet-activating factor receptor from guinea-pig lung. Nature 1991; 349:342-346.

126. Nakamura M, Honda Z, Izumi T et al. Molecular cloning and expression of platelet-activating factor receptor from human leukocytes. J Biol Chem 1991; 266: 20400-20405.

127. O'Brien PJ, Zatz M. Acylation of bovine rhodopsin by [$^3$H]palmitic acid. J Biol Chem 1984; 259:5054-5057.

128. Ovchinnikov YA, Abdulaev NG, Bogachuk AS. Two adjacent cysteine residues in the C-terminal cytoplasmic fragment of bovine rhodopsin are palmitylated. FEBS Lett 1988; 230:1-5.

129. Kennedy ME, Limbird LE. Mutations of the $\alpha_{2A}$-adrenergic receptor that eliminate detectable palmitoylation do not perturb receptor-G-protein coupling. J Biol Chem 1993; 268:8003-8011.

130. O'Dowd BF, Hnatowich M, Caron MG et al. Palmitoylation of the human $\beta_2$-adrenergic receptor. Mutation of Cys$_{341}$ in the carboxyl tail leads to an uncoupled nonpalmitoylated form of the receptor. J Biol Chem 1989; 264:7564-7569.

131. Strittmatter SM, Valenzuela D, Kennedy TE et al. G$_o$ is a major growth cone protein subject to regulation by GAP-43. Nature 1990; 344:836-841.

132. Morrison DF, O'Brien PJ, Pepperberg DR. Depalmitylation with hydroxylamine alters the functional properties of rhodopsin. J Biol Chem 1991; 266:20118-20123.

133. Karnik SS, Ridge KD, Bhattacharya S et al. Palmitoylation of bovine opsin and its cysteine mutants in COS cells. Proc Natl Acad Sci USA 1993; 90:40-44.

134. Moffett S, Mouillac B, Bonin H et al. Altered phosphorylation and desensitization patterns of a human $\beta_2$-adrenergic receptor lacking the palmitoylated Cys341. EMBO J 1993; 12:349-356.

135. van Koppen CJ, Nathanson NM. The cysteine residue in the carboxyl-terminal domain of the m2 muscarinic acetylcholine receptor is not required for receptor-mediated inhibition of adenylate cyclase. J Neurochem 1991; 57:1873-1877.

136. Karnik SS, Sakmar TP, Khorana HG. Cysteine residues 110 and 187 are essential for the formation of correct structure in bovine rhodopsin. Proc Natl Acad Sci USA 1988; 85:8459-8463.

137. Noda K, Saad Y, Graham RM et al. The high affinity state of the $\beta_2$-adrenergic receptor requires unique interaction between conserved and nonconserved extracellular loop cysteines. J Biol Chem 1994; 269: 6743-6752.

138. Curtis CAM, Wheatley M, Bansal S et al. Propylbenzilylcholine mustard labels an acidic residue in transmembrane helix 3 of the muscarinic receptor. J Biol Chem 1989;

264:489-495.

139. Kurtenbach E, Curtis CAM, Pedder EK et al. Muscarinic acetylcholine receptors. Peptide sequencing identifies residues involved in antagonist binding and disulfide bond formation. J Biol Chem 1990; 265: 13702-13708.

140. Nathans J, Davenport CM, Maumenee IH et al. Molecular genetics of human blue cone monochromacy. Science 1989; 245:831-838.

141. Lameh J, Cone RI, Maeda S et al. Structure and function of G protein coupled receptors. Pharm Res 1990; 7:1213-1221.

142. Probst WC, Snyder LA, Schuster DI et al. Sequence alignment of the G-protein coupled receptor superfamily. DNA Cell Biol 1992; 11:1-20.

143. Hedin KE, Duerson K, Clapham DE. Specificity of receptor-G protein interaction: searching for the structure behind the signal. Cell Signal 1993; 5:505-518.

144. Kühn H, Hargrave PA. Light-induced binding of guanosine triphosphate to bovine photoreceptor membranes: effect of limited proteolysis of the membranes. Biochemistry 1981; 20:2410-2417.

145. Weiss ER, Kelleher DJ, Johnson GL. Mapping sites of interaction between rhodopsin and transducin using rhodopsin antipeptide antibodies. J Biol Chem 1988; 263: 6150-6154.

146. König B, Arendt A, McDowell JH et al. Three cytoplasmic loops of rhodopsin interact with transducin. Proc Natl Acad Sci USA 1989; 86:6878-6882.

147. Francke RR, Sakmar TP, Oprian DD et al. A single amino acid substitution in rhodopsin (Lys248-leucine) prevents activation of transducin. J Biol Chem 1988; 263: 2119-2122.

148. Franke RR, König B, Sakmar TP et al. Rhodopsin mutants that bind but fail to activate transducin. Science 1990; 250: 123-125.

149. Franke RR, Sakmar TP, Graham RM et al. Structure and function in rhodopsin. Studies of the interaction between the rhodopsin cytoplasmic domain and transducin. J Biol Chem 1992; 267:14767-14774.

150. Strader CD, Dixon RAF, Cheung AH et al. Mutations that uncouple the β-adrenergic receptor from $G_s$ and increase agonist affinity. J Biol Chem 1987; 262:16439-16443.

151. O'Dowd BF, Hnatowich M, Regan JW et al. Site-directed mutagenesis of the cytoplasmic domains of the human β-adrenergic receptor. Localization of regions invlved in G protein-receptor coupling. J Biol Chem 1988; 263:15985-15992.

152. Moro O, Lameh J, Hogger P et al. Hydrophobic amino acid in the i2 loop plays a key role in receptor-G protein coupling. J Biol Chem 1993; 268:22273-22276.

153. Dixon RAF, Sigal IS, Strader CD. Structure-function analysis of the β-adrenergic receptor. Cold Spring Harbor Symp Quant Biol 1988; 53:487-498.

154. Fraser CM, Chung FZ, Wang CD et al. Site-directed mutagenesis of human β-adrenergic receptors: substitution of aspartic acid-130 by asparagine produces a receptor with high-affinity binding that is uncoupled from adenylate cyclase. Proc Natl Acad Sci USA 1988; 85:5478-5482.

155. Fraser CM, Wang CD, Robinson DA et al. Site-directed mutagenesis of m1 muscarinic acetylcholine receptors: conserved aspartic acids play important roles in receptor function. Mol Pharmacol 1989; 36:840-847.

156. Wang CD, Buck MA, Fraser CM. Site-directed mutagenesis of α₂-adrenergic receptors: identification of amino acids involved in ligand binding and receptor activation by agonists. Mol Pharmacol 1991; 40:168-179.

157. Suprenant A, Horstman DA, Akbarali H et al. A point mutation of the α₂-adrenoceptor that blocks coupling to potassium but not calcium currents. Science 1992; 257:977-980.

158. Ohyama K, Yamano Y, Chaki S et al. Domains for G-protein coupling in angiotensin II receptor type I: studies by site-directed mutagenesis. Biochem Biophys Res Commun 1992; 189:677-683.

159. Wang Z, Hang H, Ascoli M. Mutation of a highly conserved acidic residue present in the second intracellular loop of G-protein-coupled receptors does not impair hormone binding or signal transduction of the luteinizing hormone/chorionic gonadotropin receptor. Mol Endocrinol 1993; 7:85-93.

160. Dalman HM, Neubig RR. Two peptides from the $\alpha_{2A}$-adrenergic receptor alter receptor G protein coupling by distinct mechanisms. J Biol Chem 1991; 266: 11025-11029.

161. Chazenbalk GD, Nagayama Y, Russo D et al. Functional analysis of the cytoplasmic domains of the human thyrotropin receptor by site-directed mutagenesis. J Biol Chem 1990; 265:20970-20975.

162. Kosugi S, Mori T. The first cytoplasmic loop of the thyrotropin receptor is important for phosphoinositide signaling but not for agonist-induced adenylate cyclase activation. FEBS Lett 1994; 341:162-166.

163. Bommakanti RJ, Klotz K-N, Dratz EA et al. A carboxy-terminal tail peptide of neutrophil chemotactic receptor disrupts its physical complex with G protein. J Leukoc Biol 1993; 54:572-577.

164. Prossnitz ER, Quehenberger O, Cochrane CG et al. The role of the third intracellular loop of the neutrophil N-formyl peptide receptor in G protein coupling. Biochem J 1993; 294:581-587.

165. Schreiber RE, Prossnitz ER, Ye RD et al. Domains of the human neutrophil N-formyl peptide receptor involved in G protein coupling. Mapping with receptor-derived peptides. J Biol Chem 1994; 269:326-331.

166. Spengler D, Waeber C, Pantaloni C et al. Differential signal transduction of five splice variants of the PACAP receptor. Nature 1993; 365:170-175.

167. Vanetti M, Vogt G, Höllt V. The two isoforms of the mouse somatostatin receptor (mSSTR2A and mSSTR2B) differ in coupling efficiency to adenylate cyclase and in agonist-induced receptor desensitization. FEBS Lett 1993; 331:260-266.

168. Sugimoto U, Negishi M, Hayashi Y et al. Two isoforms of the $EP_3$ receptor with different carboxyl-terminal domains. Identical ligand binding properties and different coupling with $G_i$ proteins. J Biol Chem 1993; 268:2712-2718.

169. Namba T, Sugimoto Y, Negishi M et al. Alternative splicing of C-terminal tail of prostaglandin E receptor subtype EP3 determines G-protein specificity. Nature 1993; 365:166-170.

170. Wong SK-F, Parker EM, Ross EM. Chimeric muscarinic cholinergic:β-adrenergic receptors that activate $G_s$ in response to muscarinic agonists. J Biol Chem 1990; 265:6219-6224.

171. Cheung AH, Huang R-RC, Graziano MP et al. Specific activation of $G_s$ by synthetic peptides corresponding to an intracellular loop of the β-adrenergic receptor. FEBS Lett 1991; 279:277-280.

172. Cotecchia S, Ostrowski J, Kjelsberg MA et al. Discrete amino acid sequences of the $\alpha_1$-adrenergic receptor determine the selectivity of coupling to phosphatidylinositol hydrolysis. J Biol Chem 1992; 267:1633-1639.

173. Liggett SB, Caron MG, Lefkowitz RJ et al. Coupling of a mutated form of the β2-adrenergic receptor to $G_i$ and $G_s$. Requirement for multiple cytoplasmic domains in the coupling process. J Biol Chem 1991; 266:4816-4821.

174. Campbell PT, Hnatowich M, O'Dowd BF et al. Mutations of the human $\beta_2$-adrenergic receptor that impair coupling to $G_s$ interfere with receptor down-regulation but not sequestration. Mol Pharmacol 1991; 39:192-198.

175. Blüml K, Mutschler E, Wess J. Identification of an intracellular tyrosine residue critical for muscarinic receptor-mediated stimulation of phosphatidylinositol hydrolysis. J Biol Chem 1994; 269:402-405.

176. Pin J-P, Joly C, Heinemann SF et al. Domains involved in the specificity of G protein activation in phospholipase C-coupled metabotropic glutamate receptors. EMBO J 1994; 13:342-348.

177. Higashijima T, Burnier J, Ross EM. Regulation of $G_i$ and $G_o$ by mastoparan, related amphiphilic peptides and hydrophobic amines: mechanism and structural determinants of activity. J Biol Chem 1990; 265:14176-14186.

178. Nishimoto I, Hata Y, Ogata E et al. Insulin-like growth factor II stimulates calcium influx in competent BALB/c 3T3 cells primed with epidermal growth factor. Characterisitcs of calcium influx and involvement of GTP-binding protein. J Biol Chem 1987; 262:12120-12126.

179. Okamoto T, Katada T, Muruyama Y et al.

A simple structure encodes G protein-activating function of the IGFII/mannose 6-phosphate receptor. Cell 1990; 62:709-717.

180. Takahashi K, Murayama Y, Okamoto T et al. Conversion of G-protein specificity of insulin-like growth factor II/mannose-6-phosphate receptor by exchanging of a short region of β-adrenergic receptor. Proc Natl Acad Sci USA 1993; 90:11772-11776.

181. Voss T, Wallner E, Czernilofsky AP et al. Amphipathic α-helical structure does not predict the ability of receptor-derived synthetic peptides to interact with guanine nucleotide-binding regulatory proteins. J Biol Chem 1993; 268:4637-4642.

182. Cheung AH, Huang R-RC, Strader CD. Involvement of specific hydrophobic, but not hydrophilic, amino acids in the third intracellular loop of the β-adrenergic receptor in the activation of $G_s$. Mol Pharmacol 1992; 41:1061-1065.

183. Arden JR, Nagata O, Shockley MS et al. Mutational analysis of third cytoplasmic loop domains in G-protein coupling of the HM1 muscarinic receptor. Biochem Biophys Res Commun 1992; 188:1111-1115.

184. Duerson K, Carroll R, Clapham D. α-Helical distorting substitutions disrupt coupling between m3 muscarinic receptor and G proteins. FEBS Lett 1993; 324:103-108.

185. Wang H, Jaquette J, Collison K et al. Positive charges in a putative amphiphilic helix in the carboxy-terminal region of the third intracellular loop of the luteinizing hormone/chorionic gonadotropin receptor are not required for hormone-stimulated cAMP production but are necessary for expression of the receptor at the plasma membrane. Mol Endocrinol 1993; 7:1437-1444.

186. Kosugi S, Okajima F, Ban T et al. Mutation of alanine 623 in the third cytoplasmic loop of the rat thyrotropin (TSH) receptor results in a loss in the phosphoinositide but not cAMP signal induced by TSH and receptor autoantibodies. J Biol Chem 1992; 267:24153-24156.

187. Parma J, Duprez L, Van Sande J et al. Somatic mutations in the thyrotropin receptor gene cause hyperfunctioning thyroid adenomas. Nature 1993; 365:649-651.

188. Kjelsberg MA, Cotecchia S, Ostrowski J et al. Constitutive activation of the $\alpha_{1B}$-adrenergic receptor by all amino acid substitutions at a single site. J Biol Chem 1992; 267:1430-1433.

189. Ren Q, Kurose H, Lefkowitz RJ et al. Constitutively active mutants of the $\alpha_2$-adrenergic receptor. J Biol Chem 1993; 268:16483-16487. Erratum J Biol Chem 269:1566.

190. Samama P, Cotecchia S, Costa T et al. A mutation-induced activated state of the $\beta_2$-adrenergic receptor. Extending the ternary complex model. J Biol Chem 1993; 268:4625-4636.

191. Pei G, Samama P, Lohse M et al. A constitutively active mutant $\beta_2$-adrenergic receptor is constitutively desensitized and phosphorylated. Proc Natl Acad Sci USA 1994; 91:2699-2702.

192. Lefkowitz RJ, Cotecchia S, Samama P et al. Constitutive activity of receptors coupled to guanine nucleotide regulatory proteins. Trends Pharmacol Sci 1993; 14:303-307.

193. Boone C, Davis NG, Sprague GF. Mutations that alter the third cytoplasmic loop of the a-factor receptor lead to a constitutive and hypersensitive phenotype. Proc Natl Acad Sci USA 1993; 90:9921-9925.

194. Lefkowitz RJ. G protein-coupled receptor kinases. Cell 1993; 74:409-412.

195. Inglese J, Freedman NJ, Koch WJ et al. Structure and mechanism of the G protein-coupled receptor kinases. J Biol Chem 1993; 268:23735-23738.

196. Haribaru B, Snyderman R. Identification of additional members of human G-protein-coupled receptor kinase multigene family. Proc Natl Acad Sci USA 1993; 90: 9398-9402.

197. Huganir RL, Greengard P. Regulation of neurotransmitter receptor desensitization by protein phosphorylation. Neuron 1990; 5:555-567.

198. Hausdorff WP, Caron MG, Lefkowitz RJ. Turning off the signal: desensitization of β-adrenergic receptor function. FASEB J 1990; 4:2881-2889. Erratum FASEB J 4:3049.

199. Hausdorff WP, Sung J, Caron MG et al. Recent molecular analyses of β-adrenergic receptor phosphorylation, sequestration and down regulation. Asia Pacific J Pharmacol

1992; 7:149-158.

200. Lohse MJ. Molecular mechanisms of membrane receptor desensitization. Biochim Biophys Acta 1993; 1179:171-188.

201. Palczewski K, Benovic JL. G protein-coupled receptor kinases. Trends Biochem Sci 1991; 16:387-391.

202. Ohguro H, Palczewski K, Ericsson LH et al. Sequential phosphorylation of rhodopsin at multiple sites. Biochemistry 1993; 32:5718-5724.

203. Onorato JJ, Palczewski K, Regan JW et al. Role of acidic amino acids in peptide substrates of the β-adrenergic receptor kinase and rhodopsin kinase. Biochemistry 1991; 30:5118-5125.

204. Hausdorff WP, Campbell PT, Ostrowski J et al. A small region of the β-adrenergic receptor is selectively involved in its rapid regulation. Proc Natl Acad Sci USA 1991; 88:2979-2983.

205. Liggett SB, Ostrowski J, Chesnut LC et al. Sites in the third intracellular loop of the $\alpha_{2A}$-adrenergic receptor confer short term agonist-promoted desensitization. Evidence for a receptor kinase-mediated mechanism. J Biol Chem 1992; 267:4740-4746.

206. Haga T, Haga K, Kameyama K et al. Phosphorylation of muscarinic receptors: regulation by G proteins. Life Sci 1993; 52: 421-428.

207. Feramisco JR, Glass DB, Krebs EG. Optimal spatial requirements for the location of basic residues in peptide substrates for the cAMP-dependent protein kinase. J Biol Chem 1980; 255:4240-4245.

208. Hausdorff WP, Bouvier M, O'Dowd BF et al. Phosphorylation sites on two domains of the $\beta_2$-adrenergic receptor are involved in distinct pathways of receptor desensitization. J Biol Chem 1989; 264:12657-12665.

209. Clark RB, Friedman J, Dixon RAF et al. Identification of a specific site required for rapid heterologous desensitization of the β-adrenergic receptor by cAMP-dependent protein kinase. Mol Pharmacol 1989; 36:343-348.

210. Nantel F, Bonin H, Emorine LZ et al. The human $\beta_3$-adrenergic receptor is resistant to short-term agonist-promoted desensitization. Mol Pharmacol 1993; 43:548-555.

211. Clark RB, Kunkel MW, Friedman J et al. Activation of cAMP-dependent protein kinase is required for heterologous desensitization of adenylyl cyclase in S49 wild-type lymphoma cells. Proc Natl Acad Sci USA 1988; 85:1442-1446.

212. Bates MD, Caron MG, Raymond JR. Desensitization of DA1 dopamine receptors coupled to adenylyl cyclase in opossum kidney cells. Am J Physiol 1991; 260:F937-F945.

213. Lohse MJ, Benovic JL, Codina J et al. β-arrestin: a protein that regulates β-adrenergic receptor function. Science 1990; 248:1547-1550.

214. Lohse MJ, Andexinger S, Pitcher J et al. Receptor-specific desensitization with purified proteins. Kinase dependence and receptor specificity of β-arrestin and arrestin in the $\beta_2$-adrenergic receptor and rhodopsin systems. J Biol Chem 1992; 267:8558-8564.

215. Roth NS, Campbell PT, Caron MG et al. Comparative rates of desensitization of β-adrenergic receptors by the β-adrenergic receptor kinase and the cyclic AMP-dependent kinase. Proc Natl Acad Sci USA 1991; 88:6201-6204.

216. Kwatra MM, Schwinn DA, Schreurs J et al. The substance P receptor, which couples to $G_{q/11}$, is a substrate of β-adrenergic receptor kinase 1 and 2. J Biol Chem 1993; 268:9161-9164.

217. Ramkumar V, Kwatra M, Benovic JL et al. Functional consequences of $A_1$ adenosine-receptor phosphorylation by the β-adrenergic receptor kinase. Biochim Biophys Acta 1993; 1179:89-97.

218. Ishii K, Chen J, Ishii WJ et al. Inhibition of thrombin receptor signaling by a G-protein coupled receptor kinase. Functional specificity among G-protein coupled receptor kinases. J Biol Chem 1994; 269: 1125-1130.

219. Zhou X-M, Fishman PH. Desensitization of the human $\beta_1$-adrenergic receptor. Involvement of the cyclic AMP-dependent but not a receptor-specific kinase. J Biol Chem 1991; 266:7462-7468.

220. Sanchez-Yague J, Hiplin RW, Ascoli M. Biochemical properties of the agonist-induced desensitization of the follicle-stimu-

lating hormone and luteinizing hormone/chorionic gonadotropin-responsive adenylyl cyclase in cells expressing the recombinant gonadotropin receptors. Endocrinology 1993; 132:1007-1016.

221. Hipkin RW, Sánchez-Yagüe J, Ascoli M. Agonist-induced phosphorylation of the luteinizing hormone/chorionic gonadotropin receptor expressed in a stably transfected cell line. Mol Endocrinol 1993; 7:823-832.

222. Nagayama Y, Rapoport B. The thyrotropin receptor 25 years after its discovery: new insight after its molecular cloning. Mol Endocrinol 1992; 6:145-156.

223. Wojcikiewicz RJH, Tobin AB, Nahorski SR. Desensitization of cell signalling mediated by phosphoinositidase C. Trends Pharmacol Sci 1993; 14:279-285.

224. Didsbury JR, Uhing RJ, Tomhave E et al. Receptor class desensitization of leukocyte chemoattractant receptors. Proc Natl Acad Sci USA 1991; 88:11564-11568.

225. Shigemoto R, Yokota Y, Tsuchida K et al. Cloning and expression of a rat neuromedin K receptor cDNA. J Biol Chem 1990; 265:623-628.

226. Kishimoto A, Nishiyama K, Nakanishi H et al. Studies on the phosphorylation of myelin basic protein by protein kinase C and adenosine 3':5'-monophosphate-dependent protein kinase. J Biol Chem 1985; 260:12492-12499.

227. Johnson JA, Clark RB, Friedman J et al. Identification of a specific domain in the β-adrenergic receptor required for phorbol ester-induced inhibition of catecholamine-stimulated adenylyl cyclase. Mol Pharmacol 1990; 38:289-293.

228. Bouvier M, Guilbault N, Bonin H. Phorbol-ester-induced phosphorylation of the β₂-adrenergic receptor decreases its coupling to G$_s$. FEBS Lett 1991; 279:243-248.

229. Newton AC, Williams DS. Does protein kinase C play a role in rhodopsin desensitization. Trends Biochem Sci 1993; 18:275-277.

230. Ali H, Richardson RM, Tomhave ED et al. Differences in phosphorylation of formylpeptide and C5a chemoattractant receptors correlate with differences in desensitization. J Biol Chem 1993; 268:

24247-24254.

231. Von Zastrow M, Kobilka BK. Ligand-regulated internalization and recycling of human β₂-adrenergic receptors between the plasma membrane and endosomes containing transferrin receptors. J Biol Chem 1992; 267:3530-3538.

232. Tan PK, Davis NG, Sprague GF et al. Clathrin facilitates the internalization of 7 transmembrane segment receptors for mating pheromones in yeast. J Cell Biol 1993; 123:1707-1716.

233. Sibley DR, Strasser RH, Benovic JL et al. Phosphorylation/dephosphorylation of the β-adrenergic receptor regulates its functional coupling to adenylate cyclase and subcellular distribution. Proc Natl Acad Sci USA 1986; 83:9408-9412.

234. Waldo GL, Northup JK, Perkins JP et al. Characterization of an altered membrane form of the β-adrenergic receptor produced during agonist-induced desensitization. J Biol Chem 1983; 258:13900-13908.

235. Yu SS, Lefkowitz RJ, Hausdorff WP. β-adrenergic receptor sequestration. A potential mechanism of receptor desensitization. J Biol Chem 1993; 268:337-341.

236. Lameh J, Philip M, Sharma Y et al. Hm1 muscarinic cholinergic receptor internalization requires a domain in the third cytoplamic loop. J Biol Chem 1992; 267:13406-13412.

237. Moro O, Shockley MS, Lameh J et al. Overlapping multi-site domains of the muscarinic cholinergic Hm1 receptor involved in signal transduction and sequestration. J Biol Chem 1994; 269:6651-6655.

238. Strader CD, Sigal IS, Blake AD et al. The carboxyl terminus of the hamster β-adrenergic receptor expressed in mouse L cells is not required for receptor sequestration. Cell 1987; 49:855-863.

239. Cheung AH, Dixon RAF, Hill WS et al. Separation of the structural requirements for agonist-promoted activation and sequestration of the β-adrenergic receptor. Mol Pharmacol 1990; 37:775-779.

240. Maeda S, Lameh J, Mallet WG et al. Internalization of the Hm1 muscarinic cholinergic receptor involves the third cytoplasmic loop. FEBS Lett 1990; 269:386-388.

241. Benya RV, Fathi Z, Battey JF et al. Serines and threonines in the gastrin-releasing peptide receptor carboxyl terminus mediate internalization. J Biol Chem 1993; 268: 20285-20290.

242. Bouvier M, Hausdorff WP, De Blasi A et al. Removal of phosphorylation sites from the $\beta_2$-adrenergic receptor delays onset of agonist-promoted desensitization. Nature 1988; 333:370-373.

243. Nussenzveig DR, Heinflink M, Gershengorn MC. Agonist-stimulated internalization of the thyrotropin-releasing hormone receptor is dependent on two domains in the receptor carboxyl terminus. J Biol Chem 1993; 268:2389-2392.

244. Moro O, Lameh J, Sadée W. Serine-and threonine-rich domain regulates internalization of muscarinic cholinergic receptors. J Biol Chem 1993; 268:6862-6865.

245. Barak LS, Tiberi M, Freedman NJ et al. A highly conserved tyrosine residue in G protein-coupled receptors is required for agonist-mediated $\beta_2$-adrenergic receptor sequestration. J Biol Chem 1994; 269:2790-2975.

246. Bouvier M, Collins S, O'Dowd BF et al. Two distinct pathways for cAMP-mediated down-regulation of the $\beta_2$-adrenergic receptor. Phosphorylation of the receptor and regulation of its mRNA level. J Biol Chem 1989; 264:16786-16792.

247. Mahan LC, Koachman AM, Insel PA. Genetic analysis of $\beta$-adrenergic receptor internalization and down-regulation. Proc Natl Acad Sci USA 1985; 82:129-133.

248. Hadcock JR, Ros M, Malbon CC. Agonist regulation of $\beta$-adrenergic receptor mRNA. Analysis in S49 mouse lymphoma mutants. J Biol Chem 1989; 264:13956-13961.

249. Yang J, Logsdon CD, Johansen TE et al. Human m3 muscarinic acetylcholine receptor carboxyl-terminal threonine residues are required for agonist-induced receptor down-regulation. Mol Pharmacol 1993; 44: 1158-1164.

250. Valiquette M, Bonin H, Hnatowich M et al. Involvement of tyrosine residues located in the carboxyl tail of the human $\beta$-adrenergic receptor in agonist-induced down-regulation of the receptor. Proc Natl Acad Sci USA 1990; 87:5089-5093.

251. Valiquette M, Bonin H, Bouvier M. Mutation of tyrosine-350 impairs the coupling of the $\beta_2$-adrenergic receptor to the stimulatory guanine nucleotide binding protein without interfering with receptor downregulation. Biochemistry 1993; 32: 4979-4985.

252. Wang H-Y, Berrios M, Malbon CC. Localization of $\beta$-adrenergic receptors in A341 cells *in situ*. Effect of chronic exposure to agonist. Biochem J 1989; 263:533-538.

253. Zemcik BA, Strader CD. Fluorescent localization of the $\beta$-adrenergic receptor on DDT-1 cells. Down-regulation by adrenergic agonists. Biochem J 1988; 251:333-339.

254. Wang S-Z, Hu JR, Long RM et al. Agonist-induced down-regulation of m1 muscarinic receptors and reduction of their mRNA level in a transfected cell line. FEBS Lett 1990; 276:185-188.

255. Fukamauchi F, Hough C, Chuang DM. Expression and agonist-induced down-regulation of mRNAs of m2- and m3-muscarinic acetylcholine receptors in cultured cerebellar granule cells. J Neurochem 1991; 56:716-719.

256. Habecker BA, Nathanson NM. Regulation of muscarinic acetylcholine receptor mRNA expression by activation of homologous and heterologous receptors. Proc Natl Acad Sci USA 1992; 89:5035-5038.

257. Fujimoto CS, Straub RE, Gershergorn MC. Thyrotropin-releasing hormone (TRH) and phorbol myristate acetate decrease TRH receptor messenger RNA in rat pituitary GH3 cells: evidence that protein kinase-C mediates the TRH effect. Mol Endocrinol 1991; 5:1527-1532.

258. Akamizu T, Ikuyama S, Saji M et al. Cloning, chromosomal assignment, and regulation of the rat thyrotropin receptor: expression of the gene is regulated by thyrotropin, agents that increase cAMP levels, and thyroid autoantibodies. Proc Natl Acad Sci USA 1990; 87:5677-5681.

259. Hadcock JR, Malbon CC. Down-regulation of $\beta$-adrenergic receptors: agonist-induced reduction in receptor mRNA levels. Proc Natl Acad Sci USA 1988; 85:5021-5025.

260. Hadcock JR, Wang H, Malbon CC. Agonist-induced destabilization of $\beta$-adrenergic

receptors mRNA. Attenuation of glucocorticoid induced upregulation of β-adrenergic receptors. J Biol Chem 1989; 19928-19933.

261. Lu DL, Peegel H, Mosier SM et al. Loss of lutropin/human choriogonadotropin receptor messenger ribonucleic acid during ligand-induced down-regulation occurs posttranscriptionally. Endocrinology 1993; 132:235-240.

262. Fujimoto J, Narayanan CS, Benjamin JE et al. Mechanism of regulation of thyrotropin-releasing hormone receptor messenger ribonucleic acid in stably transfected rat pituitary cells. Endocrinology 1992; 130:1879-1884.

263. Port JD, Huang LY, Malbon CC. β-adrenergic agonists that down-regulate receptor mRNA up-regulate a M(r) 35,000 protein(s) that selectively binds to β-adrenergic receptor mRNAs. J Biol Chem 1992; 267:24103-24108.

264. Narayanan CS, Fujimoto J, Geras-Raaka E et al. Regulation by thyrotropin-releasing hormone (TRH) of TRH receptor mRNA degradation in rat pituitary GH3 cells. J Biol Chem 1992; 267:17296-17303.

265. Hadcock JR, Malbon CC. Regulation of receptor expression by agonists: transcriptional and post-transcriptional controls. Trends Neurosci 1991; 14:242-247.

266. Collins S, Caron MG, Lefkowitz RJ. From ligand binding to gene expression: new insights into the regulation of G-protein-coupled receptors. Trends Biochem Sci 1992; 17:37-39.

# CHAPTER 4

# G Protein-Coupled Receptors, Pathology and Therapeutic Strategies

The identification of molecular defects underlying a variety of disease states has seen great advances in recent years with the application of molecular biological approaches to investigation of clinical samples and in chromosomal mapping studies and genetic linkage analysis. Given the widespread distribution of G protein-coupled receptors and the diversity of ligands with which they interact, discrete receptors may be considered candidates for involvement in a number of pathological conditions.

## MOLECULAR DEFECTS IN G PROTEIN-COUPLED RECEPTORS: PHENOTYPIC VARIATION AND INVOLVEMENT IN PATHOLOGICAL CONDITIONS

### 1. RHODOPSIN AND THE VISUAL COLOR PIGMENTS: RETINITIS PIGMENTOSA AND OTHER VISUAL DISORDERS

#### i. Retinitis pigmentosa

The term retinitis pigmentosa (RP) refers to a genetically and clinically heterogeneous group of human inherited degenerative retinopathies. The disease derives its name from the formation of visible pigmentory patches in the retina, resulting from invasion by cells of the retinal pigmented epithelium. RP is characterized by loss of photoreceptor function, which is manifested as night blindness and progressive loss of peripheral vision and visual acuity.[1] Clinical heterogeneity is observed in both the time course and regional distribution of photoreceptor cell degeneration.[2]

RP represents the most common inherited form of human visual handicap, with an estimated prevalence of approximately 1 in 3000. The genetic heterogeneity of the disease is highlighted by transmission in an autosomal dominant, autosomal recessive or X-linked fashion. The involvement of rhodopsin, the visual pigment of the rod photoreceptor, in the pathogenesis of RP was established through localization of one of the autosomal dominant RP genes to chromosome $3_q$,[3] the same chromosomal location as the gene encoding rhodopsin (see chapter 1, Table 1.2). Subsequently, a range of mutations in rhodopsin has been shown to cosegregate with autosomal dominant RP in different families, accounting for approximately 30% of cases of autosomal dominant RP.[1]

The amino acid residues of rhodopsin which are the site of mutations identified in autosomal dominant RP are depicted in Figure 4.1 and specific codon changes are listed in Table 4.1.[1,4-6] To date, 35 amino acid substitutions, 2 deletion mutations, 2 premature termination mutations and 1 mutation of a consensus splice donor site have been detected in autosomal dominant RP. The majority of the substitution mutations involve amino acid residues which are invariant among the visual pigments (see Fig. 4.1). The location of the mutation within the consensus splice site sequence would be anticipated to result in production of a molecule potentially lacking the C-terminal segment.[4] An additional termination mutation, representing mutation of codon 249, which encodes Glu in the wild-type rhodopsin molecule, has been identified in autosomal recessive RP (see Fig. 4.1).[7]

Mechanisms underlying the phenotypic variability of the termination mutations remain to be elucidated. However, the functional significance of a range of mutations observed in autosomal dominant RP has been examined by generation of mutant rhodopsin molecules and determination of their properties in vitro in transfected mammalian cells.

A substantial number of the mutations that occur in transmembrane (TM) domains involve the replacement or introduction of a charged amino acid residue. Consistent with predictions that such changes could influence structural properties of the molecule such as folding or disposition within the membrane, molecules carrying the mutations Thr58Arg, Val87Asp, Gly89Asp, Arg135Leu and Arg135Trp, have been shown to exhibit lower levels of accumulation, variable regeneration with the chromophore 11-*cis*-retinal and inefficient translocation from the endoplasmic reticulum to the plasma

membrane.[8] The mutations Arg135Leu and Arg135Trp represent mutation of the invariant Arg residue of the highly conserved (Asp/Glu-Arg-Tyr) sequence which occurs in the majority of G protein-coupled receptors (see chapter 1, Fig. 1.5). Association of a pathological condition with mutation of this residue highlights an important role for this sequence in maintaining the integrity of members of the superfamily. Disruption of the conformation of α-helical TM segments would also be expected with replacement of Pro residues at positions 53, 171, and 267, or the introduction of a Pro residue at position 211, as is observed in mutations of the rhodopsin molecule giving rise to autosomal dominant RP. Two of these mutations, Pro171Leu and Pro267Leu, occurring in TM helices IV and VI, respectively, involve Pro residues highly conserved among members of the receptor superfamily (see chapter 1, Fig. 1.5).

Defective accumulation and regeneration kinetics are also associated with the Tyr178Cys mutant[8] and may be predicted for other mutations clustered within the second intradiscal or extracellular loop (see Fig. 4.1 and Table 4.1). It has been suggested that substitutions within the second intradiscal loop could adversely affect protein folding by affecting the apposition of Cys residues 110 and 187, which form an essential intramolecular disulfide bond.[4]

The wild-type rhodopsin molecule exhibits two potential sites for N-linked glycosylation within the N-terminal intradiscal domain. The functional significance of these sites is highlighted by identification of the mutation Thr17Met, which destroys one of these sites, underlying one form of autosomal dominant RP.

The identification of relatively fewer mutations occurring on the cytoplasmic face of the rhodopsin molecule (see Fig. 4.1; Table 4.1) may reflect a lesser importance for the three cytoplasmic loops and C-terminal segment of the molecule in maintaining the three-dimensional structure of the protein.[4] Consistent with this, no functional deficit has been observed in

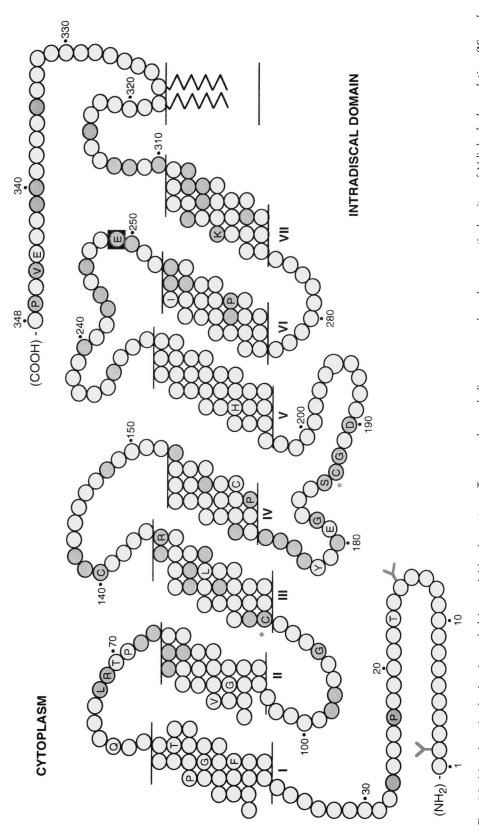

Fig. 4.1. Mutations in rhodopsin underlying retinitis pigmentosa. Transmembrane helices are numbered consecutively, sites of N-linked glycosylation (Y) and palmitoylation are indicated, and Cys residues (Cys110 and Cys187), which are involved in disulfide bond formation, are identified with an asterisk. Amino acid residues that are invariant among vertebrate opsins are shown in blue and residues mutated in autosomal dominant retinitis pigmentosa are identified. Identification of the residue Glu249 (boxed) represents the mutation Glu249ter that underlies one form of autosomal recessive retinitis pigmentosa. [Modified with permission from American Journal of Human Genetics Volume 53, Macke JP, Davenport CM, Jacobson SG et al. Identification of novel rhodopsin mutations responsible for retinitis pigmentosa: implications for the structure and function of rhodopsin, Pages 80-89, Copyright (1993)].

either the Gln344ter or Pro347Leu mutants which have been examined in transfected cells.[8] Mechanisms underlying the involvement of such mutations of the rhodopsin molecule in the pathogenesis of RP remain to be defined.

The amino acid residues Lys296 and Glu113 are of critical importance in the rhodopsin molecule, as they provide, respectively, the site of attachment of 11-*cis*-retinal to the protein through a protonated Schiff base linkage and the Schiff base

*Table 4.1. Rhodopsin mutations in autosomal dominant retinitis pigmentosa*

| Codon | Sequence change | Mutation |
|---|---|---|
| 17 | ACG → ATG | Thr → Met |
| 23 | CCC → CAC | Pro → His |
| 23 | CCC → CTC | Pro → Leu |
| 45 | TTT → CTT | Phe → Leu |
| 51 | GGC → GTC | Gly → Val |
| 53 | CCC → CGC | Pro → Arg |
| 58 | ACG → AGG | Thr → Arg |
| 64 | CAG → TAG | Gln → ter |
| Δ68-71 | ΔCTGCGCACGCCT | del LeuArgThrPro |
| 87 | GTC → GAC | Val → Asp |
| 89 | GGT → GAT | Gly → Asp |
| 106 | GGG → TGG | Gly → Trp |
| 106 | GGG → AGG | Gly → Arg |
| 125 | CTG → CGG | Leu → Arg |
| 135 | CGG → CTT | Arg → Leu |
| 135 | CGG → GGG | Arg → Gly |
| 135 | CGG → TGG | Arg → Trp |
| 140 | TGT → TCT | Cys → Ser |
| 167 | TGC → CGC | Cys → Arg |
| 171 | CCA → CTA | Pro → Leu |
| 178 | TAC → TGC | Tyr → Cys |
| 181 | GAC → AAG | Glu → Lys |
| 182 | GGC → AGC | Gly → Ser |
| 186 | TCG → CCG | Ser → Pro |
| 188 | GGA → AGA | Gly → Arg |
| 188 | GGA → GAA | Gly → Glu |
| 190 | GAC → AAC | Asp → Asn |
| 190 | GAC → GGC | Asp → Gly |
| 211 | CAC → CCC | His → Pro |
| 211 | CAC → CGC | His → Arg |
| Δ255 | ΔATC | del Ile |
| 267 | CCC → CTC | Pro → Leu |
| 296 | AAG → GAG | Lys → Glu |
| 296 | AAG → ATG | Lys → Met |
| 344 | CAG → TAG | Glu → ter |
| 345 | GTG → ATG | Val → Met |
| 347 | CCG → TCG | Pro → Ser |
| 347 | CCG → CTG | Pro → Leu |
| 347 | CCG → CGG | Pro → Arg |
|  | $G_{4335}$ → T | Consensus splice donor site |

Abbreviations:  del  deletion
ter  termination codon
Information in this table is derived from Lindsay et al. (1992),[1] Macke et al. (1993),[4] Sullivan et al. (1993),[5] and Rao et al. (1994).[6]

counterion, while the residue His211 is believed to play a role in conformational changes associated with activation of the rhodopsin molecule (see chapter 3). It has been demonstrated that the autosomal dominant RP mutants Lys296Glu and Lys296Met exhibit constitutive activation, with activation of the rhodopsin-associated G protein transducin occurring in either the absence of chromophore or the absence of light. It is possible that photoreceptor cell degeneration in individuals carrying this mutation may result from persistent stimulation of the phototransduction pathway.[5,9] The development of retinal derivatives with the capacity to inactivate specifically such constitutively active mutant molecules represents a promising avenue for therapeutic intervention.[10] Aberrant transport or membrane destabilization by mutant proteins may be responsible for photoreceptor cell degeneration in individuals carrying other mutations in the rhodopsin molecule, but details of the degenerative process remain to be defined. Transgenic animal technology has recently been applied successfully to generate a suitable murine model of the human disease condition attributable to the Pro23His mutation in rhodopsin.[11] Continuing utilization of such methodology should provide the means to address specific questions relating to mechanisms associated with progression of the disease.

## ii. Congenital stationary night blindness

The prominent symptom of congenital stationary night blindness is night blindness, a symptom also observed in RP. However, in contrast to RP, this form of night blindness is not associated with an apparent degeneration of rod or cone photoreceptors. The disease is also genetically heterogeneous, with transmission occurring by autosomal dominant, autosomal recessive and X-linked means.

Two constitutively active mutant rhodopsin molecules, comprising the mutation Gly90Asp within TM helix II or Ala292Glu within TM helix VII of the protein, have been identified in individuals with this disease.[6,12] The topographical organization of TM helices of rhodopsin places the Asp90 and Glu292 mutations in close proximity to Glu113 and Lys296, the two residues normally involved in formation of a salt bridge linkage with the chromosphore retinal. As in autosomal dominant RP, constitutive activity of the photoreceptor, leading to night blindness, arises from disruption of this salt bridge. However, there are phenotypic differences between the constitutively active Lys296Glu and Lys296Met mutants which occur in RP and the Gly90Asp and Ala292Glu mutants detected in congenital stationary night blindness. These differences may be attributed to the level of activation of the different molecules. Activation of the Gly90Asp and Ala292Glu mutants will occur independently of light but, in contrast to the Lys296Glu and Lys296Met mutants, requires the presence of the chromophore 11-*cis*-retinal. Since the chromophore is in dynamic equilibrium with the rhodopsin molecule and in continuous exchange within photoreceptor outer segments, the level of activation of the Gly90Asp and Ala292Glu mutant rhodopsin molecules would presumably be less intense than the constitutively activated state of the Lys296Glu and Lys296Met mutants which give rise to RP. This lower level of activation is presumed to represent a less toxic condition.[6,12]

## iii. Congenital blue cone monochromacy and tritanopia

Blue cone monochromacy is a rare X-linked disorder of color vision characterized by the absence of both red and green cone sensitivities. It arises from alterations in the red and green visual pigment (opsin) gene cluster (see chapter 1, Table 1.2). Two classes of alteration have been observed in families carrying the disease. In one class of mutation, sequences upstream of the red and green pigment genes, which are essential for expression of the genes, are deleted. In the other class of mutation, affected individuals carry a red-green hybrid opsin gene, in which there is a point mutation resulting in the substitution Cys203Arg. This point mutation results in loss of blue-green dichromacy that characterizes inheritance of the red-green hybrid opsin gene. Since the residues Cys126 and Cys203 of red and green opsins correspond to residues Cys110 and Cys187 of the rhodopsin molecule (see Fig. 4.1) which are involved in the formation of an essential intramolecular disulfide bond, mutation of Cys203 in the hybrid red-green opsin molecule is believed to result in the formation of a nonfunctional pigment protein due to interference with the structural integrity of the protein.[13]

Tritanopia is an autosomal dominant genetic disorder of human vision characterized by a selective deficiency of blue spectral sensitivity. Two different point mutations, each leading to amino acid substitutions in the blue opsin protein, have been identified in individuals with this disease. The substitutions Gly79Arg or Ser214Pro, occurring in TM helices II and V, respectively, are nonconservative and might each be expected to perturb the folding, processing or stability of the mutant proteins.[14]

## 2. Vasopressin V₂ Receptor: Nephrogenic Diabetes Insipidus

Nephrogenic diabetes insipidus (NDI) is usually an X-linked disorder, although there have been reports of patterns of inheritance that are not X-linked[15,16] and an autosomal form of NDI has been described in the mouse.[17] NDI is characterized by insensitivity of the distal nephron of the kidney to the antidiuretic action of the pituitary hormone vasopressin. Affected individuals present early in neonatal life with excessive water intake (polydipsia), excessive urine excretion (polyuria) and failure to concentrate urine. They exhibit dehydration, fever, anorexia and vomiting. Central nervous system (CNS) complications may develop due to dehydration of the brain. Following precise localization of the gene responsible for NDI to $X_{q28}$ using genetic linkage analysis,[18] it was demonstrated that

derivatives of somatic cell hybrids which carry the human gene locus for NDI express functional vasopressin $V_2$ receptors.[19] Coincidence of the physical map location of the vasopressin $V_2$ receptor with the locus for NDI was established subsequently by molecular cloning and chromosomal localization of the vasopressin $V_2$ receptor gene (see chapter 1, Table 1.2).[20-22]

The amino acid residues of the vasopressin $V_2$ receptor that have been identified in patients with NDI as sites of mutation and cosegregate with the disease are depicted in Figure 4.2 and specific codon changes are listed in Table 4.2. To date, ten nonconservative amino acid substitutions, two deletion mutations, one premature termination mutation and three frameshift mutations have been identified.[23-29] The frameshift mutations, occurring at Ile228 in TM5 or after Arg230 or Gly246, in the third cytoplasmic loop of the receptor (see Fig. 4.2), would result in incorporation of an additional 29, 39 or 23 amino acids, respectively, of nonreceptor sequence before reaching a termination codon.[25] The premature termination mutation would give rise to a more severely truncated receptor protein, with truncation occurring within TM3. This mutation was detected in the most severely affected patient and would presumably result in a completely nonfunctional protein.[25]

The normal vasopressin $V_2$ receptor contains a single Cys residue within each of the first and second extracellular loops. These residues correspond to Cys residues which are highly conserved in G protein-coupled receptors and are involved in the formation of an intramolecular disulfide bond that is critical for maintaining the structural integrity of a number of receptors (see chapter 1, Figs. 1.4 and 1.5). Consequently, it is highly likely that the first and second extracellular loops of the normal vasopressin $V_2$ receptor are constrained through the formation of a disulfide bond. It has also been suggested that these domains may be spatially contiguous and contribute to the vasopressin binding site.[30] Four of the amino acid substitution

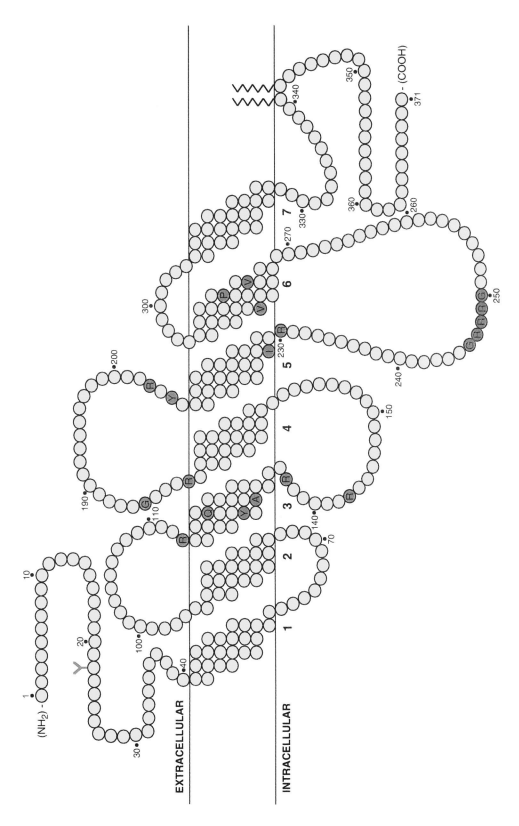

Fig. 4.2. Mutations in the vasopressin $V_2$ receptor underlying nephrogenic diabetes insipidus. Transmembrane helices are numbered consecutively and potential sites for N-linked glycosylation (Y) and palmitoylation are indicated. Amino acid residues mutated in nephrogenic diabetes insipidus are identified.

**Table 4.2. *Vasopressin V$_2$ receptor mutations in nephrogenic diabetes insipidus***

| Codon | Sequence change | Mutation |
|---|---|---|
| 113 | CGG → TGG | Arg → Trp |
| 119 | CAG → TAG | Gln → ter |
| 128 | TAC → TCC | Tyr → Ser |
| 132 | GCC → GAC | Ala → Asp |
| 137 | CGC → CAC | Arg → His |
| 143 | CGT → CCT | Arg → Pro |
| 181 | GGC → TGC | Arg → Cys[a] |
| 185 | GGT → TGT | Gly → Cys |
| 203 | CGC → TGC | Arg → Cys |
| 205 | TAT → TGT | Tyr → Cys |
| 228 | ATC → ACT | Insertion → Frameshift |
| 230 | CCGG → GG | Deletion → Frameshift |
| 246 | GGG → GG | Deletion → Frameshift |
| Δ247-250 | ΔCGCCGCAGGGGA | del ArgArgArgGly[a] |
| Δ278 or 279 | ΔGTC | del Val |
| 286 | CCC → CGC | Pro → Arg |

Abbreviations:    del   deletion
                  ter   termination codon
[a] One of two mutations detected in one patient and three affected family members
Information in this table is derived from Rosenthal et al. (1992),[23] van den Ouweland et al. (1992),[24] Pan et al. (1992),[25] Holtzman et al. (1993),[26] Merendino et al. (1993),[27] Rosenthal et al. (1993),[28] and Tsukaguchi et al. (1993).[29]

mutations detected in NDI involve the introduction of an additional Cys residue within the second extracellular loop of the receptor (see Fig. 4.2 and Table 4.2). This additional Cys residue could potentially interfere with correct disulfide bonding between the first and second extracellular loops, leading to inappropriate protein folding and structural instability or disruption of the vasopressin binding site. One of the Cys substitution mutations, Arg181Cys, was detected in a patient who also exhibited deletion of four amino acids within the third cytoplasmic loop (see Fig. 4.2 and Table 4.2). This deletion occurs in a region of the vasopressin V$_2$ receptor which is highly divergent in sequence between human and rat receptors, leading to the suggestion that the deletion mutation may have little effect on receptor function and the primary defect in the mutated receptor in this family results from the Cys substitution mutation.[24,25]

Two additional mutations associated with NDI involve amino acid residues which are highly conserved in G protein-coupled receptors. The mutation Arg137His occurs at the interface of TM3 with the second intracellular loop (see Fig. 4.2) and involves the invariant Arg residue of the highly conserved sequence (Asp/Glu-Arg-Tyr) (see chapter 1, Fig. 1.5). Vasopressin V$_2$ receptors carrying this mutation exhibit no alteration in ligand binding properties compared to wild-type receptors, but lack the ability to stimulate adenylyl cyclase activity, consistent with demonstration of the importance of this intracellular region for G protein coupling in other receptors (see chapter 3).[28] The mutation Pro286Arg, within TM7 (see Fig. 4.2), results in substitution of a conserved Pro residue (see chapter 1, Fig. 1.5) which would normally function as an α-helix breaker. It has been suggested that this conserved residue may play a role in the creation of a ligand binding pocket within the membrane, raising the possibility that vasopressin $V_2$ receptors carrying the Pro286Arg mutation may have altered ligand binding capability.[25]

The mutation Ala132Asp occurs within TM3 (see Fig. 4.2) and represents a nonconservative amino acid substitution which would alter the hydropathic profile of the receptor. As has been demonstrated for the Val87Asp and Gly89Asp mutations of the rhodopsin molecule, which also involve the substitution of a charged residue for an uncharged polar or nonpolar residue (see Section 1(i), above), such mutant receptors would be expected to have altered structural properties. However, the identification of specific defects in receptor function and pathogenetic consequences attributable to this mutation and to other identified mutations in the vasopressin $V_2$ receptor awaits examination of the properties of mutated receptors both in vitro in transfected mammalian cells and in transgenic animal model systems.

## 3. MSH RECEPTOR: SKIN PIGMENTATION

The genetic locus determining pigmentation phenotype in a number of mammalian species, including the dog, horse, leopard and mouse, is termed the "extension" locus. It exhibits dominant and recessive alleles, with dark coat colors being due to dominant alleles and recessive alleles giving rise to light coat colors. The molecular basis of allelic differences at the murine extension locus has been elucidated with identification of the extension locus as encoding the melanocyte-stimulating hormone (MSH) MC-1 receptor and characterization of MSH receptors expressed in phenotypically different mice.

The recessive allele "yellow" ($e$) results from a frameshift mutation between TM domains 5 and 6, that produces a truncated nonfunctional receptor protein which extends for 12 amino acids before terminating. Dominant alleles, on the other hand, arise from point mutations within the MSH receptor which result in receptor hyperactivity. For example, the dominant allele "tobacco darkening" ($E^{tob}$) encodes a receptor with a single point mutation of C to T, resulting in the amino acid substitution Ser69Leu within the first intracellular loop. This mutant receptor responds to MSH stimulation by greater activation of adenylyl cyclase than the level of activation exhibited by the wild-type allele. Similarly, two "sombre" alleles, $E^{so}$ and $E^{so-3J}$, each arise from a single point mutation. The allele $E^{so}$ is attributable to the substitution of C for T, which results in the amino acid substitution Leu98Pro within the first extracellular loop. The allele $E^{so-3J}$ represents the substitution of G for A that gives rise to the amino acid substitution Glu92Lys within TM2, and results in a receptor which is no longer responsive to hormone but is constitutively active and capable of eliciting the generation of high levels of intracellular cAMP in the absence of agonist.[31]

## 4. LH/CG RECEPTOR: FAMILIAL MALE PRECOCIOUS PUBERTY

Familial male precocious puberty (FMPP) is an autosomal dominant, male-limited disorder characterized by the onset of puberty as early as age four. The pathological condition reflects unregulated Leydig cell activity, with production of testosterone and hyperplasia of Leydig cells in the testes occurring at prepubertal levels of circulating luteinizing hormone (LH). Linkage of FMPP to a single point mutation within the luteinizing hormone/chorionic gonadotropin (LH/CG) receptor has identified the molecular defect underlying this disease. A single base change from A to G, resulting in the mutation Asp578Gly within TM6 of the receptor, generates a receptor which remains responsive to LH but also exhibits functional activation in the absence of agonist.

The pathophysiology of FMPP may be attributed to constitutive activation of the LH/CG receptor-mediated cAMP signaling pathway. In normal circumstances, the production of testosterone and initiation of steroidogenesis in Leydig cells involves elevation of intracellular cAMP levels due to the activation of LH/CG receptors in response to increased circulating levels of LH. Leydig cell hyperfunction and hyperplasia may arise from the expression of constitutively active mutant LH/CG receptors, which are capable of the generation of markedly increased levels of intracellular cAMP independently of LH stimulation. The identification of constitutively active LH/CG receptors in boys affected by FMPP also provides an explanation for restriction of this condition to boys. Inappropriate activation of LH/CG receptors could be responsible for precocious puberty in boys since LH is sufficient to trigger steroidogenesis in Leydig cells, but would not lead to precocious puberty in girls since both LH and follicle-stimulating hormone (FSH) are required for activation of ovarian steroidogenesis.[32]

## 5. ACTH RECEPTOR: FAMILIAL GLUCOCORTICOID DEFICIENCY

Familial glucocorticoid deficiency is a rare autosomal recessive disorder also known as hereditary unresponsiveness to adrenocorticotropic hormone (ACTH). Affected children usually have recurrent hypoglycemic episodes or convulsions, and exhibit progressive hyperpigmentation, weakness, failure to thrive and excessively frequent and severe infections in early childhood. Death may occur in the first two years of life.

Discrete mutations in the ACTH receptor have recently been identified in two affected individuals. One affected individual was shown to be homozygous for a nonconservative substitution, Ser74Ile, within TM2 of the receptor.[33] This mutation results in approximately 10-fold reduced responsiveness of the receptor to ACTH,[34] and involves an amino acid residue conserved across the melanocortin receptor family, which includes receptors for ACTH and melanocyte-stimulating hormone (MSH) (see chapter 1, Table 1.1).[35,36] While the role of this conserved Ser residue in this receptor family is not known, the activation properties of ACTH receptors carrying the Ser74Ile mutation are

consistent with a role for this residue in ligand binding.[34]

A second affected individual has been shown to be heterozygous for mutation of the ACTH receptor, with each ACTH receptor allele carrying a discrete point mutation. One mutant receptor allele exhibits mutation of the codon for Arg201 to a termination codon, giving rise to a receptor molecule prematurely terminated after TM5, which would presumably be nonfunctional. The second mutant receptor allele carries the mutation Ser120Arg in TM3. Such mutant ACTH receptors would be expected to have altered structural properties due to the nonconservative substitution of a charged residue into the hydrophobic TM environment. A contribution of both mutant receptor alleles to the disease phenotype in this case is indicated, since parents and grandparents of the affected individual, each carrying only one mutant ACTH receptor allele and reporting no symptoms or signs of adrenal dysfunction, were shown to have subclinical resistance to ACTH.[37]

## 6. Ca²⁺-SENSING RECEPTOR: FAMILIAL HYPOCALCIURIC HYPERCALCEMIA AND NEONATAL SEVERE HYPERPARATHYROIDISM

Familial hypocalciuric hypercalcemia (FHH) and neonatal severe hyperparathyroidism (NSHPT) are two inherited conditions characterized by altered calcium homeostasis. FHH is an autosomal dominant condition characterized by moderate elevation of serum $Ca^{2+}$ and low excretion of urinary $Ca^{2+}$, while levels of parathyroid hormone (PTH), which is secreted by the parathyroid gland to control the level of serum $Ca^{2+}$, remain inappropriately normal. This condition is also known as familial benign hypercalcemia because the usual clinical sequelae of hypercalcemia, including altered mental status, kidney stones, decreased urinary concentrating ability and hypertension, do not occur. NSHPT is the homozygous form of FHH. It is associated with marked elevation of both serum $Ca^{2+}$ and PTH, and is generally lethal if total parathyroidectomy is not performed early in life.[38,39]

Following correlation of the localization of the gene defect responsible for FHH to chromosome $3_{q2}$ with mapping of the human homolog of the recently described bovine parathyroid $Ca^{2+}$-sensing receptor to chromosome 3, three discrete mutations in the human $Ca^{2+}$-sensing receptor which segregate with disease phenotype have been identified. These comprise the nonconservative amino acid substitutions Arg186Glu and Glu298Arg in the N-terminal extracellular domain of the receptor and Arg796Trp within the membrane-proximal segment of the third intracellular loop at the interface with TM5. These mutations occur in residues which are conserved not only between human and bovine $Ca^{2+}$-sensing receptor sequences,[40] but also between these receptors and other members of the metabotropic glutamate receptor subfamily (see chapter 1, Fig. 1.6). A markedly blunted response to $Ca^{2+}$ characterizes the $Ca^{2+}$-sensing receptor carrying the mutation Arg796Trp, while the Arg186Glu and Glu298Arg mutations represent alteration of conserved charge and might be predicted to affect the structure of the N-terminal extracellular domain, potentially causing interference with receptor expression or ligand binding properties.[39]

# G PROTEIN-COUPLED RECEPTORS IN DEVELOPMENT, MALIGNANT TRANSFORMATION, VIRAL INFECTION AND AUTOIMMUNE DISEASE

The molecular characterization of a large number of G protein-coupled receptors has allowed examination of sites of receptor expression in normal and pathological conditions and assessment of the functional consequences of receptor activation. Such investigations have provided insight into interacting systems which operate during normal development, as well as identification of mechanisms associated with normal and malignant proliferative processes, viral infection and the functioning of receptors which are targets of autoimmune responses.

## 1. DEVELOPMENT

### i. Fertilization and gamete activation

The process of fertilization relies upon reciprocal spermatozoal and oocyte activation events. Oocyte-induced sperm activation allows sperm to penetrate the *zona pellucida* (ZP), the extracellular matrix of the oocyte, and fuse with the oocyte plasma membrane. Ensuing sperm-induced oocyte activation involves a series of "early" and "late" events which lead to DNA synthesis and entry of the fertilized oocyte into the meiotic cycle.[41]

In the mouse, sperm activation is a G protein-mediated event, induced by interaction of specific sperm cell surface receptors with the oocyte-specific glycoprotein ZP3.[42] There is increasing evidence that subsequent sperm-induced oocyte activation also involves ligand-induced activation of G protein signal transduction pathways. A candidate sperm-specific membrane-associated ligand that may mediate sperm-oocyte plasma membrane binding and fusion has recently been identified in the guinea pig.[43] Early events of oocyte activation may be mimicked by exogenous application of activators of G proteins, such as GTPγS, or by the second messenger molecules inositol 1,4,5-trisphosphate ($IP_3$) or diacylglycerol (DAG) generated by G protein-coupled effector enzymes.[44-47] In addition, the full complement of early and late events of oocyte activation has been observed in response to ligand activation of human m1 muscarinic acetylcholine receptors, which couple to phosphatidylinositol (PI) hydrolysis, when these receptors are expressed in murine oocytes.[48] These data strongly suggest mediation of sperm-induced oocyte activation by G proteins expressed in the oocyte and highlight the importance of PI metabolism in this process. They also indicate the existence of specific oocyte cell surface receptor(s) which would presumably comprise as yet unidentified members of the G protein-coupled receptor superfamily.

## ii. GHRH receptor: dwarfism

Normal growth and development relies critically upon regulation of the release of growth hormone from somatotrophs in the anterior pituitary gland. The hypothalamic peptides somatostatin and growth hormone-releasing hormone (GHRH) have complementary biological roles in the hypothalamic-pituitary axis. They mediate the inhibition and stimulation, respectively, of growth hormone secretion, through interaction with specific G protein-coupled receptors on somatotrophs.[49] The critical role of GHRH receptors in normal growth processes is highlighted by the identification of either aberrant expression or function of these receptors underlying the discrete murine dwarf phenotypes "snell" and "little", respectively.

The GHRH receptor is normally expressed on nascent somatotroph precursor cells in the developing pituitary gland. Its activation in response to GHRH released from fibers of the median eminence results in elevation of intracellular cAMP levels, cAMP-dependent gene transcription and cellular proliferation.[50-53] Snell (dw/dw) dwarf mice carry a mutation in the gene encoding Pit-1, a tissue-specific POU-domain transcription factor which is normally expressed within the developing pituitary gland in somatotroph precursor cells[54,55] and is responsible for the initial activation of the growth hormone gene. These dwarf mice are unable to express growth hormone and exhibit pituitary hypoplasia and absence of somatotroph, lactotroph and thyrotroph cells in the anterior pituitary gland.[56-59] An absence of GHRH receptor transcripts in nascent somatotrophs correlates with the proliferative defect in these mice, identifying a role for GHRH receptor in the coordinate control of proliferation and differentiation of somatotrophs in the developing anterior pituitary gland.[60] Whether Pit-1 activates expression of the GHRH receptor gene directly or indirectly remains to be established. However, molecular cloning of the GHRH receptor[60,61] provides a means for elucidation of mechanisms regulating receptor expression and function.

A second murine dwarf phenotype, referred to as "little" (lit/lit), is also characterized by anterior pituitary hypoplasia, as well as reduced levels of expression of growth hormone and prolactin.[62,63] This phenotype arises from a single point mutation, resulting in the nonconservative amino acid substitution Asp60Gly, in the N-terminal extracellular domain of the GHRH receptor. GHRH receptors carrying this mutation are unable to elicit elevation of intracellular cAMP levels required for transcriptional and proliferative events associated with normal somatotroph development.[64,65] The amino acid residue Asp60 of the GHRH receptor corresponds to Asp71 of the rat secretin receptor[66] and represents a residue which is conserved in all members of the secretin subfamily of G protein-coupled receptors described to date (see chapter 1, Fig. 1.7). Identification of aberrant function in receptors carrying a nonconservative substitution of this residue may reflect its crucial importance for structural properties of the secretin receptor subfamily.

## iii. TSH receptor: hypothyroidism

The hypothyroid (hyt/hyt) mouse phenotype, characterized by severe hypothyroidism, thyroid hyporesponsiveness and elevated thyroid-stimulating hormone (TSH) levels, is an autosomal recessive condition with fetal onset that represents a model of human sporadic congenital hypothyroidism. Hypothyroidism arises from dysfunction of TSH receptor-mediated stimulation of adenylyl cyclase activity in the thyroid gland and in the congenitally hypothyroid mouse is attributable to a single point mutation in the TSH receptor. The amino acid substitution Pro556Leu in TM4 of the TSH receptor, while not affecting the level of expression of mutant receptor mRNA, results in abolition of TSH binding and functional receptor activation. This mutation represents substitution of a Pro residue conserved in the majority of G protein-coupled receptors (see chapter 1, Fig. 1.5) and corresponds to the Pro171Leu mutation in TM helix IV of rhodopsin that gives rise to one form of autosomal dominant RP (See Fig. 4.1 and Table 4.1). Pathological conditions associated with mutation of conserved Pro residues highlight their importance in maintenance of the structural integrity of receptors in the superfamily.[67]

## 2. MALIGNANT TRANSFORMATION

The potential of G protein-coupled receptors to act as protooncogenes was initially suggested indirectly by the demonstration that the protein encoded by the *mas* oncogene is a member of the receptor superfamily.[68] The human *mas* gene product was subsequently shown to be tumorigenic when transfected into murine NIH 3T3 fibroblasts.[69] It has been suggested from in vitro analyses that the *mas* oncogene may either encode a functional angiotensin II receptor or may respond to an unidentified ligand to enhance the responsiveness of endogenous angiotensin II signaling systems occurring in certain cell types, resulting in activation of PI signaling pathways and mitogenesis.[70,71]

Transforming activity in vitro has been reported for other G protein-coupled receptors which couple to PI metabolism. The muscarinic acetylcholine receptor subtypes m1, m3 and m5 act as agonist-dependent oncogenes and exhibit dose-dependent transforming activity when transfected into NIH 3T3 fibroblasts. The m2 and m4 receptor subtypes, on the other hand, which couple to inhibition of adenylyl cyclase activity, are unable to elicit transformation of transfected cells.[72] Similarly, the serotonin 5-HT$_{2C}$, previously designated 5-HT$_{1C}$, receptor subtype[73] triggers serotonin-dependent malignant transformation of transfected murine NIH 3T3 and BALB/c-3T3 fibroblasts.[74,75] Cells derived from foci of NIH 3T3 cells transformed with the 5-HT$_{2C}$ receptor will give rise to tumors in vivo in nude mice.[74]

Considerable accumulating evidence for an involvement of PI hydrolysis in mitogenic processes[76] has led to examination of the signaling pathways mediating this response, and correlation with mechanisms previously identified in growth factor and oncogene signal transduction. For example, the mitogenic effect of the peptides bombesin, vasopressin and endothelin-1, acting through endogenous G protein-coupled receptors on murine Swiss 3T3 fibroblasts, is associated with stimulation of protein tyrosine kinase activity.[77] Similarly, mitogenesis resulting from the activation of endogenous endothelin receptors on glomerular mesangial cells involves autophosphorylation of pp60$^{c-src}$ protein tyrosine kinase activity and pp60$^{c-src}$-catalysed phosphorylation of a specific peptide substrate. This indicates cross-talk between components of signaling pathways associated with G protein-coupled receptors and nonreceptor protein tyrosine kinases, such as pp60$^{c-src}$[78] The activation of transforming G protein-coupled receptors, such as the m1 muscarinic acetylcholine receptor transfected into NIH 3T3 fibroblasts, also results in tyrosine phosphorylation of cellular proteins, and it has been suggested that growth promoting pathways activated by receptors coupled to G proteins may involve tyrosine phosphorylation of a subset of proteins previously identified as substrates for oncogene-encoded tyrosine kinases.[79] The reactivation of DNA synthesis induced by agonist stimulation of the muscarinic acetylcholine m1 receptor transfected into NIH 3T3 fibroblasts occurs independently of adenylyl cyclase and the conventional protein kinase C, (PKC)-α. However, it does involve the Ca$^{2+}$-independent PKC-ζ, which has been identified as a critical component of mitogenic signal transduction.[80,81]

A role for mutations in G protein-coupled receptors in tumor induction and/or progression was first demonstrated with the α$_{1B}$-adrenergic receptor. When transfected into Rat 1 or murine NIH 3T3 fibroblasts, the α$_{1B}$-adrenergic receptor will induce the formation of foci in confluent cell monolayers in an agonist-dependent manner. Mutational alteration of the receptor in the C-terminal segment of the third cytoplasmic loop with three amino acids substitutions, Arg288Lys, Lys290His and

Ala293Leu (Fig. 4.3a), renders the receptor constitutively active. This is manifested as an enhanced ability to induce transformation. Focus formation induced by the mutated receptor is no longer dependent on agonist stimulation, but can be augmented by the administration of agonist.[82] Similarly, two somatic mutations in the TSH receptor, also in the C-terminal segment of the third cytoplasmic loop, have recently been identified in tumor tissue isolated from patients with hyperfunctioning thyroid adenoma. The mutations comprise the single amino acid substitutions Asp619Gly, observed in two patients, or Ala623Ile, observed in one patient (Fig. 4.3b). In a manner analogous to the mutant construct of the $\alpha_{1B}$-adrenergic receptor, these naturally-occurring mutant TSH receptors remain responsive to TSH but exhibit constitutive activation. This results in elevated adenylyl cyclase activity and approximately two- to three-fold higher basal accumulation of cAMP. Expression of these mutant receptors in thyrocytes, whose function and growth is stimulated by cAMP levels, results in induction of hyperfunctioning adenoma.[83]

The neuropeptide bombesin and related mammalian neuropeptides gastrin-releasing peptide (GRP) and neuromedin B, as well as bradykinin, cholecystokinin (CCK), galanin, neurotensin and vasopressin, have been reported to stimulate clonal growth in a number of human small cell lung cancer (SCLC) cell lines.[84-86] Some of these peptides, including CCK, GRP, neurotensin and vasopressin, are secreted by SCLC cells and by some SCLC tumors[87] and hence could act as autocrine or paracrine mitogens. Other peptides with reported mitogenic activity, such as bradykinin and galanin, are not produced by SCLC cells. All of the peptides capable of stimulating clonal growth couple to intracellular PI hydrolysis[73] and have been shown to cause mobilization of intracellular $Ca^{2+}$.[86]

The mitogenic effects of bombesin and the related peptides GRP and neuromedin B on SCLC clonal cell lines in vitro and on SCLC xenografts in vivo in nude mice may be inhibited by a monoclonal antibody directed against bombesin-like peptides.[88] The use of specific antibodies to clear secreted mitogenic factors represents a possible therapeutic strategy which may see application in particular cases. In addition, synthetic analogs of bombesin, as well as a number of analogs of substance P, which act as broad spectrum peptide receptor antagonists, will also inhibit clonal growth of SCLC cell lines in vitro and the growth of SCLC xenografts in vivo in nude mice.[85,89-91] The mechanisms of action of some of these antagonists remain to be defined, but their identification represents a promising avenue of investigation for the development of novel therapeutic agents of low antigenicity and high tissue penetration.

Overexpression of members of the G protein-coupled receptor superfamily has been reported in a variety of tumors. For example, overexpression of transcripts for the $CCK_A$ and novel expression of $CCK_B$ receptor subtypes is a prominent feature of azaserine-induced rat pancreatic carcinoma.[92] Similarly, somatostatin receptors represent a consistent pathobiochemical marker for adrenal and extra-adrenal pheochromocytomas and paragangliomas and have been reported to occur in some primary breast cancers. While the significance of receptor expression remains unclear, the expression of receptors at detectable levels clearly represents potential for the development of diagnostic protocols for in vivo localization of tumors or detection of recurrent disease.[93,94]

## 3. Viral Infection

Genes encoding putative G protein-coupled receptors have been described in the genomes of three mammalian viruses, comprising human cytomegalovirus (HCMV), herpesvirus saimiri (HVS), which infects primates other than man, and swinepox virus (see chapter 1, Table 1.1).

HCMV infects leukocytes, fibroblasts and epithelial cells and may cause multisystem and fatal disease depending upon the timing of acquisition and immunocompetence of the host. The HCMV genome contains three predicted open reading frames, UL27, UL28 and UL33, which exhibit many of the sequence features conserved in the majority of G protein-coupled receptors (see chapter 1, Fig. 1.5) and bear a low degree of homology (20-30%) to a variety of mammalian G protein-coupled receptors including neurokinin A (NKA; substance K), muscarinic acetylcholine and $\beta_2$-adrenergic receptors.[95] The role of these putative receptor sequences in the life cycle of the virus remains unknown. However, the UL28 gene product has been demonstrated to bind the β chemokines macrophage inflammatory proteins (MIP)-1α and -1β, human monocyte chemoattractant protein (MCP)-1 and "regulated on activation, normal T expressed and secreted" (RANTES),[96] leading to the suggestion that expression of this virally-encoded chemokine binding protein during a viral infection may interfere with pro-inflammatory responses elicited by activation of host chemokine receptors. The term "molecular piracy" has been coined to describe this proposed strategy of eluding host defenses through molecular mimicry of proteins normally involved in host defense mechanisms.[97] The demonstration of specific ligand binding activity also identifies at least the UL28 receptor sequence as a potential target for the development of antiviral therapeutics. Given the homology exhibited by these sequences with other known mammalian G protein-coupled receptors, it is possible that existing pharmacological agents which interact with receptors such as adrenergic and muscarinic acetylcholine receptors may be useful in therapeutic application to control the pathogenesis of HCMV infection.

Herpesvirus saimiri (HVS) attacks T-lymphocytes and causes fatal lymphoproliferative diseases, including leukemias and lymphomas, in a number of nonhuman primates. The *ECRF3* gene of HVS encodes a receptor protein[98] which bears approximately 30% amino acid identity to known mammalian receptors for α and β chemokines, namely the interleukin (IL)-8 A and B receptors, and the MIP-1α/RANTES receptor, respectively. The ECRF3 receptor

protein has a ligand specificity identical to that of the IL-8 B receptor. It exhibits functional activation in response to the α chemokines IL-8, melanoma growth stimulatory activity (MGSA/GRO) and neutrophil activating protein (NAP)-2, but does not interact with β-chemokines.[97] The known growth factor activity of IL-8 and MGSA/GRO clearly suggests a role for the ECRF3 receptor in proliferation, while the novel demonstration of transmembrane signaling by a viral receptor in response to mammalian ligands suggests previously undefined roles for chemokines in the molecular pathogenesis of viral infections.

## 4. AUTOIMMUNE DISEASE

Autoimmune disease is characterized by a specific cell-mediated or humoral immune response against constituents of the body's own tissues. The self antigens, or autoantigens, may be cytoplasmic or cell surface molecules.[99] Autommimune diseases attributable to the development of antibodies which recognize cell surface receptors of both the ligand-gated ion channel and G protein-coupled receptor superfamilies have been reported. Myasthenia gravis is an organ-specific autoimmune disease caused by an antibody-mediated assault on the muscle nicotinic acetylcholine receptor, a ligand-gated ion channel, at the neuromuscular junction.[100-102] The existence of autoantibodies against G protein-coupled receptors has been reported in allergic respiratory disease and Cushing's disease, with an involvement of $\beta_2$-adrenergic and adrenocorticoptropic hormone (ACTH) receptors, respectively.[103] In the autoimmune thyroid disorders hyperthyroid Graves' disease and hypothyroid idiopathic myxedema, humoral immunity is directed against the G protein-coupled TSH receptor.[104,105]

TSH receptor autoantibodies comprise an heterogeneous population which plays a major role in the pathogenesis of thyroid disorder by influencing the function and growth of the thyroid gland. This may be manifested as stimulation of the activity of the thyroid gland, as observed in Graves' disease, which results in elevation

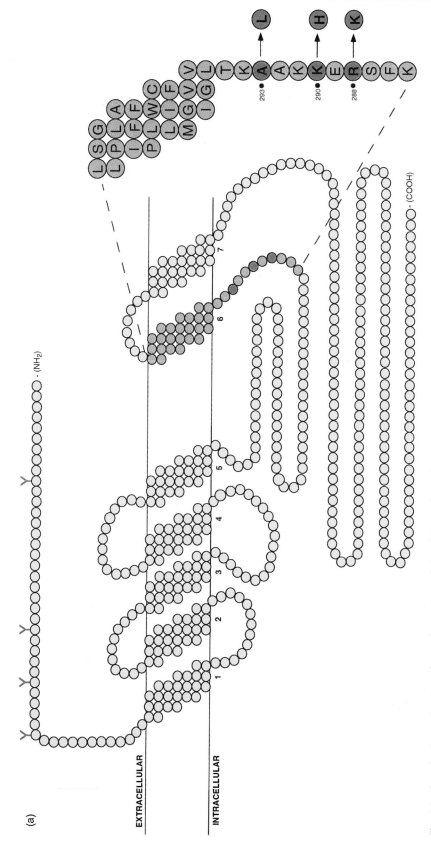

Fig. 4.3. Constitutive activation of α_{1B}-adrenergic and thyroid-stimulating hormone receptors. Model of (a) the α_{1B}-adrenergic receptor, and (b) the thyroid-stimulating hormone receptor, with transmembrane helices numbered consecutively and potential sites for N-linked glycosylation (Y) indicated. The C-terminal segment of the third intracellular loop and the sixth transmembrane domain are enlarged. Mutation of the α_{1B}-adrenergic receptor that gives rise to constitutive receptor activation and two mutations identified in patients with hyperfunctioning adenoma are shown in green. [(a) Reprinted with permission from Proceedings of the National Academy of Sciences USA Volume 88, Allen LF, Lefkowitz RJ, Caron MC et al, G-protein-coupled receptor genes as protooncogenes: constitutively activating mutation of the α_{1B}-adrenergic receptor enhances mitogenesis and tumorigenicity, Pages 11354-11358, Copyright (1991). (b) Reprinted with permission from Nature Volume 365, Parma J, Duprez L, Van Sande J et al, Somatic mutations in the thyrotropin receptor gene cause hyperfunctioning thyroid adenomas, Pages 649-651, Copyright (1993) Macmillan Magazines Limited].

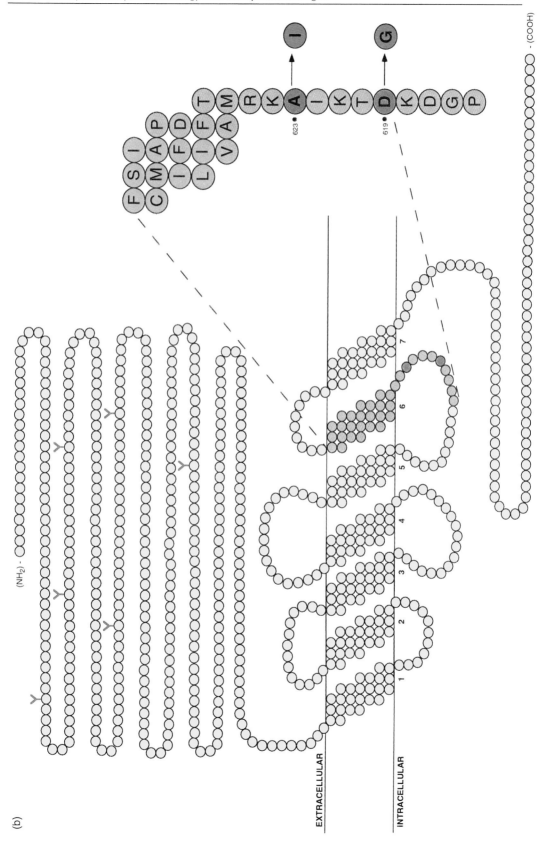

(b)

of circulating levels of thyroid hormones thyroxine ($T_4$) or triiodothyronine ($T_3$). Hypothyroidism, which is characteristic of idiopathic myxedema, arises from a different spectrum of autoantibodies which block functional activation of TSH receptor, while growth-promoting autoantibodies give rise to endemic or sporadic goiters.[105,106]

The molecular basis for the autoimmune phenotype in Graves' disease and idiopathic myxedema has been elucidated with delineation of determinants on the TSH receptor which are recognized by the two types of autoantibody involved. The TSH receptor contains an immunogenic domain with no identified functional determinants. This occurs within the N-terminal extracellular domain of the receptor, flanking TM1, and encompasses amino acid residues 303-382, with a particular involvement of residues 352-366. Autoantibodies responsible for idiopathic myxedema recognize determinants flanking this immunogenic domain and interfere with high affinity binding of TSH to its receptor at amino acid residues 295-306, 385 and 387-395 (see chapter 3). Such steric hindrance is considered to be responsible for the hypothyroid condition. The molecular determinant for antibodies involved in Graves' disease is centered on amino acid residues 38-45, also within the N-terminal extracellular domain of the TSH receptor. This epitope occurs within the region defined by residues 12-50 which are also involved in TSH binding (see chapter 3). Interaction of autoantibodies at this functional site of the receptor could result in receptor activation, which would account for the thyroid stimulatory activity of these autoantibodies.[107]

## GENETIC LINKAGE ANALYSIS

The basis for studies directed towards chromosomal localization of genes responsible for pathological conditions has been provided by the relatively recent availability of molecular probes that reveal restriction fragment length polymorphisms (RFLPs) and the identification of polymorphic microsatellite DNA sequences, combined with genetic linkage analysis with known chromosomal markers. The application of such molecular biological techniques in the analysis of the molecular genetic basis of disease has met with considerable success in elucidation of the underlying defect in conditions such as Duchenne muscular dystrophy,[108] cystic fibrosis,[109] fragile X syndrome[110] and Huntington's disease,[111] where the pathological phenotype may be attributed to defects in a single gene.

Greater difficulty has been encountered in attempts to define gene defects responsible for complex inherited diseases such as the neuropsychiatric illnesses bipolar affective disorder and schizophrenia. There may be considerable heterogeneity in disease phenotype, which could arise from an underlying genetic heterogeneity, compounded with incomplete and variable expression of genotype, cumulative effects of several mutant genes, as well as environmental factors. With neuropsychiatric and behavioral disorders, variability in diagnostic criteria may further confound definition of disease phenotype and accurate identification of affected individuals within pedigrees under investigation.

Genetic linkage analyses conducted to date on complex diseases have incorporated both genome scanning and screening with molecular probes for discrete candidate genes. Among candidate genes of potential relevance to bipolar affective disorder, schizophrenia, Gilles de la Tourette syndrome and addictive states, including alcoholism, are the genes encoding several G protein-coupled neurotransmitter receptors.

### 1. BIPOLAR AFFECTIVE DISORDER

Bipolar affective disorder, also termed manic depressive illness, is a disorder of mood characterized by recurrent periods of mania and major depression. During periods of mania, patients exhibit expansiveness, elation, agitation, hyperexcitability, hyperactivity and apparent increased speed of thought and speech. Depressive episodes are characterized by feelings of sadness, despair and discouragement, and are often associated with feelings of low self-esteem, guilt and self-reproach, withdrawal from inter-personal contact and somatic symptoms such as eating and sleep disturbances.[99]

Available evidence from family, twin and adoption studies clearly indicates an important role for genetic factors in the predisposition for developing bipolar affective disorder.[112-114] However, segregation analyses have been unable to define the mode of transmission of this disorder, and it is not clear whether the condition may be due to a single major gene, a small number of genes, or multifactorial polygenic inheritance. An initial report of linkage of bipolar affective disorder to genes in the chromosomal region $11_{p15}$ in a large Amish family has failed to be replicated in subsequent analyses.[115-122] To date, systematic screens of the genome have also not yielded the identity of a susceptibility locus for bipolar affective disorder.[123-129]

Identification of candidate genes which may be of relevance to bipolar affective disorder has involved examination of behavioral effects of neurotransmitter agonists, as well as assessment of the mechanisms of action of therapeutic agents such as lithium and carbamezapine which are used in the treatment of acute mania. Theoretical considerations have suggested a role for dopamine and serotonin neurotransmitter receptors, as well as for components of intracellular signaling pathways, such as G proteins. However, examination of a large number of pedigrees has, to date, excluded linkage of bipolar affective disorder to the dopamine $D_1$, $D_2$, $D_3$ and $D_4$[130-137] and serotonin 5-$HT_{1A}$[138] receptor subtypes, and to the $G_s\alpha$ subunit.[139]

## 2. SCHIZOPHRENIA

Schizophrenia is a neuropsychiatric disorder or heterogeneous group of disorders characterized by marked disorder of thought, including delusions, and visual and auditory hallucinations. It is accompanied by mood disturbances and deterioration from a previous level of functioning. Affected individuals experience an altered sense of self and relationship to the external world, which may present as loss of ego boundaries, dereistic thinking or autistic withdrawal, and they exhibit disturbances in behavior, with apparently purposeless or stereotyped activity or inactivity.[99,140] There is extensive phenotypic variation in presentation of the disorder. Family, twin and adoption studies indicate a clear genetic component, with between 66 and 93% of the variance in the etiology of schizophrenia estimated to be genetic in origin.[141-144] However, segregation analyses have not given a clear indication of the mode of transmission.[145]

The action of neuroleptic therapeutic drugs in blockade of dopaminergic neuronal transmission has led to the suggestion that overactivity of certain dopaminergic pathways within the brain may underlie the schizophrenic condition. The psychotomimetic effects of drugs such as cocaine, LSD and amphetamines, which promote the release of dopamine in vivo, lend support to this hypothesis.[146] It has been established that the density of the dopamine $D_2$-like receptor subtypes $D_2$ and $D_4$ is consistently elevated in postmortem schizophrenic brain tissue, while the density of dopamine $D_1$ receptors is normal.[147,148] Molecular cloning of dopamine receptor subtypes has allowed demonstration that therapeutic concentrations of antipsychotic drugs, such as haloperidol, chlorpromazine, remoxipride, raclopride, risperidone, sulpiride and thioridazine, inhibit the activity of dopamine $D_2$ receptors, while therapeutic concentrations of the antipsychotic drug clozapine act at the dopamine $D_4$ receptor subtype.[149] It should be noted, however, that identification of dopamine receptors as a site of action for these therapeutic drugs does not preclude their interaction with other receptors, such as the serotonin 5-$HT_{2A}$ receptor, which may result in variation in clinical efficacy in alleviation of certain symptoms.[147,150]

Despite considerable evidence for an involvement of dopaminergic signaling in the pathology of schizophrenia, genetic linkage analyses with molecular probes for all five dopamine receptor subtypes, $D_1$-$D_5$, have failed to implicate the dopamine receptor genes directly in the etiology of schizophrenia.[151-158] Similarly, the serotonin 5-$HT_{2A}$ receptor has been excluded as a major susceptibility gene for schizophrenia in at least one large kindred.[159]

DNA polymorphisms in the dopamine $D_1$ and $D_2$ receptors have been identified in postmortem schizophrenic brain tissue. In the case of the dopamine $D_1$ receptor, the DNA polymorphisms identified to date do not introduce any changes to the amino acid sequence of the receptor.[160] Of DNA polymorphisms reported within the $D_2$ receptor gene, several cause no alteration of the $D_2$ receptor sequence at the amino acid level,[161,162] while one reported polymorphism results in the single amino acid substitution Ser311Cys, within the third intracellular loop of the $D_2$ receptor. However, the frequency of occurrence of this structural variant of the dopamine $D_2$ receptor in schizophrenic brain tissue was not significantly different from its occurrence in normal brain tissue.[163] The DNA polymorphism of the dopamine $D_4$ receptor gene which gives rise to a receptor variant carrying an additional four amino acids in the N-terminal extracellular segment of the receptor (see chapter 1) does not occur in greater frequency in schizophrenic patients, relative to controls, but may be of significance in delusional disorder.[164] No allelic association of the other polymorphic variants of the dopamine $D_4$ receptor, which differ in the length of the third intracellular loop and exhibit altered ligand binding affinities (see chapter 1), has been found.[165,166] As with structural variants of the dopamine $D_3$ receptor (see chapter 1), the significance of these polymorphic variants of the dopamine $D_4$ receptor remains to be defined.

While the genetic linkage and molecular biological analyses conducted to date do not provide evidence for defects in dopamine receptor genes giving rise to schizophrenia, they do not eliminate the possibility that other gene products involved in regulation of the expression or functional activation of dopamine receptors may play a role in the etiology of schizophrenia. Further refinement of technologies such as positron emission tomography (PET), which allows investigation of regional brain energy metabolism, may become increasingly valuable in investigation of the role of specific receptor subtypes in the pathology of complex diseases such as schizophrenia.[167,168]

## 3. GILLES DE LA TOURETTE SYNDROME

Gilles de la Tourette syndrome is an hereditary, neuropsychiatric disorder characterized by motor tics, usually beginning in the face and progressing to other areas, and by vocal tics, such as hissing, snorting and barking. Some patients also exhibit behavioral problems and psychopathology including obsessionality, depression and anxiety.[169-171] The pattern of inheritance is believed to be autosomal dominant, with incomplete penetrance and variable expression.[172]

The motor and vocal tic symptoms of Gilles de la Tourette syndrome have traditionally been treated with dopamine receptor antagonists, such as haloperidol, pimozide and sulpiride, and there is evidence that obsessional symptoms are also reduced by dopamine receptor antagonists.[170,171] Clozapine is also an effective alternative in the treatment of tremor.[173] However, despite indications of the involvement of dopaminergic neuronal pathways in manifestations of the disease, genetic linkage analyses have failed to implicate the genes encoding the dopamine receptors, $D_1$-$D_5$, in the etiology of the syndrome.[172,174-177]

## 4. DRUG ADDICTION AND ALCOHOLISM

Although drug addiction has been manifest in human populations for many centuries, only recently has progress been made in elucidation of the molecular events underlying the addictive state. Nevertheless, neither the mechanisms responsible for the addictive nature of specific drugs, nor genetic factors which render certain individuals particularly vulnerable to addiction, have been clearly defined.

Addiction represents the compulsive use of a drug despite adverse consequences, and is usually associated with the phenomena of tolerance and dependence. Tolerance refers to diminished efficacy of a drug as a consequence of repeated exposure, with increased dosage required to achieve the same initial effect. Dependence represents a need for continued exposure to avoid a withdrawal syndrome, with attendant physical or psychological disturbances upon cessation of use of the drug.[178]

Since the discovery of endogenous opioid peptides and the identification of specific G protein-coupled receptors which bind opiate ligands within discrete regions of the brain, considerable research has been directed towards examination of biochemical and molecular mechanisms responsible for changes in brain function that result in tolerance, dependence and addiction to opiate drugs and cocaine. These studies have failed to generate any evidence for association of addiction with either consistent changes in opioid receptor number or affinity for ligands.[179] Nor do changes in levels of endogenous opioid peptides appear to underlie tolerance or dependence. Similarly, although cocaine and other addictive psychostimulants either acutely inhibit the reuptake of monoamine neurotransmitters or stimulate their release, no consistent long-term changes in specific neurotransmitter or receptor systems have been demonstrated in brain regions thought to underlie psychostimulant addiction.[180-182] However, some of the changes in brain function which occur in response to chronic drug administration may involve alterations in the level of expression of G protein subunits and up-regulation of the generation of intracellular cAMP and protein phosphorylation signaling pathways, as has been observed in opiate-sensitive neurons as they adapt to chronic morphine administration. Such changes in neuronal biochemistry may underlie both physical and psychological aspects of drug addiction.[178]

The recent molecular cloning of the opioid receptor subtypes (see chapter 1, Table 1.1), and in particular the μ opioid receptor, which interacts with the analgesic morphine, provides a complementary avenue of investigation. The expression of discrete populations of cloned receptors in mammalian cell lines provides the means for screening and identification of potentially novel pharmacological agents which may be less subject to the development of tolerance. Such agents would not only be of substantial value in therapeutic application. They would also be valuable reagents in continuing investigations of cellular adaptation processes that underlie addiction and may lead to the development of therapeutics which prevent or reverse the actions of addictive drugs on specific target neurons.

Susceptibility to drug and alcohol abuse is believed to be multifactorial in transmission, with involvement of biological, psychological and social factors. Family, twin and adoption studies support a genetic contribution, but the genetic basis for predisposition to the development of an addictive phenotype remains obscure. Candidate genes for involvement in substance abuse vulnerability include neurotransmitter transporters such as the dopamine transporter, which is a direct target of cocaine action.[183] However, the dopamine transporter gene has been shown by genetic linkage analysis not be associated with polysubstance abuse.[184] Reports of linkage of the dopamine $D_2$ receptor gene to alcoholism[185,186] have been contested[187-190] and the issue of association of the dopamine $D_2$ receptor with alcohol abuse remains controversial.[191-195]

The neuropsychiatric disorders and other multifactorial illnesses currently represent the most difficult areas for application of genetic techniques in elucidation of the etiology of disease. However, progress may be anticipated with a greater availability of DNA markers, more detailed chromosomal maps and information derived from the Human Genome Project. In addition, mechanization of genotype determination, implementation of work station technology and greater computational power in the analysis of genetic data will

tion, and provide the
for detailed assessment
the vast number of G
eptors and molecules
ression and function in

## RECEPTOR SUBTYPES, SUBTYPE-SPECIFIC AGENTS AND NOVEL THERAPEUTIC STRATEGIES

Classical approaches incorporating pharmacological characterization of receptor subtypes and medicinal chemistry have resulted in the development of a large number of therapeutic agents which interact with G protein-coupled receptors and exhibit receptor subtype selectivity. These include β-adrenergic receptor antagonists, which are widely used in the treatment of hypertension,[196] β-adrenergic receptor agonists used in asthma therapy[197] and histamine $H_1$ receptor antagonists administered in management of inflammatory and allergic conditions.[198] However, our understanding of the functional diversity of subtypes of receptors specific for a particular endogenous ligand is increasingly incorporating an appreciation of the structural diversity of receptor subtypes at the molecular level. The actions of many therapeutics which exhibit specificity for a pharmacologically defined receptor subtype may now be explained in molecular terms.

The ability to undertake in vitro characterization of ligand binding and activation properties of discrete receptor subtypes, together with the availability of molecular probes for determination of sites of expression of particular receptor subtypes, have given insights into the efficacy of certain drugs, such as clozapine, which is administered to schizophrenic patients. Such investigations also allow a clearer delineation of target sites and a more directed approach in drug design, as has been achieved with development of the serotonin 5-$HT_1$ receptor agonist sumatriptan for the treatment of migraine. Novel therapeutic approaches may be based on research findings into the complexity of regulation

of interacting physiological systems, and take advantage of tissue-specific expression of receptor subtypes. Such strategies may encompass the development of agents which act at receptors for neuromodulatory agents such as adenosine or the neuropeptides, whose roles in many physiological systems are only beginning to be elucidated. In addition, the availability of nucleotide sequence information for discrete receptor subtypes raises possibilities for the application of gene therapy, using antisense technology in therapeutic paradigms.

## 1. CLOZAPINE AND SCHIZOPHRENIA

Antipsychotic drugs administered to individuals afflicted with schizophrenia are known to be dopamine receptor antagonists. Most such drugs show preferential affinity for the dopamine $D_2$ receptor subtype and, in addition to their desired therapeutic effect, produce extrapyramidal side-effects, tardive dyskinesia and neuroendocrine effects. These side-effects are not observed with the "atypical" neuroleptic clozapine, which shows higher affinity for the dopamine $D_4$ than for the dopamine $D_2$ receptor subtype.[149,199]

The availability of molecular probes for the dopamine receptor subtypes has allowed demonstration of an abundance of dopamine $D_2$ receptors in the striatum, a brain structure which plays an important role in motor behavior. Dopamine $D_4$ receptor mRNA, on the other hand, is expressed in the frontal cortex, midbrain, amygdala and medulla and occurs at very low levels in the striatum. Thus, extrapyramidal side-effects associated with neuroleptic therapy may reflect blockade of striatal dopamine $D_2$ receptors, while the lack of side-effects associated with clozapine administration may arise from the differential distribution of dopamine $D_4$ receptors for which clozapine exhibits higher affinity.[200] The efficacy of clozapine in the treatment of schizophrenia clearly suggests an important role for the dopamine $D_4$ receptor subtype in psychotic conditions. However, clozapine also exhibits high affinity for other receptors, including serotonin 5-$HT_{2C}$ and

5-HT$_{2A}$, $\alpha_1$-adrenergic, muscarinic acetylcholine and histamine H$_1$ receptors.[201] The potential involvement of these receptors in schizophrenia remains to be determined.

## 2. SUMATRIPTAN AND MIGRAINE

Classical migraine is an episodic, intense, throbbing headache, often unilateral, which is frequently associated with or preceded by irritability, nausea, photophobia, phonophobia and gastrointestinal disturbances such as diarrhea or constipation. The precise mechanisms responsible for a migraine attack are unknown. However, the headache is believed to be of vascular origin, since the brain is largely insensitive to pain. Migraine attacks commence with constriction of cranial arteries, often with visual symptoms, and are then associated with a vasodilatory phase.[99,202,203]

Until relatively recently, the mainstay of acute therapy for migraine was administration of the ergot-based vasoconstrictor drugs ergotamine and dihydroergotamine. However, the clinical usefulness of these drugs is limited due to considerable side-effects, including nausea, vomiting, abdominal pain, diarrhea, muscle cramps, limb paresthesia and generalized vasoconstriction.[203,204]

The neurotransmitter serotonin has potent vasoconstrictor activity on both peripheral and cranial blood vessels, but has only relatively recently been shown to mediate its effects through the pharmacologically discrete 5-HT$_2$ and 5-HT$_1$ subtypes, respectively.[203] Delineation of differential sites of expression of distinct serotonin receptor subtypes, together with demonstration of the efficacy of slow intravenous infusion of serotonin in alleviation of migraine headache,[205,206] provided a rational basis for development of a selective serotonin receptor agonist with selectivity for cranial vasculature that may be effective in the treatment of migraine. Sumatriptan (3-[2-(dimethylamino)ethyl]-N-methyl-IH-indole-5-methane sulfonamide) (GR43175) has subsequently been shown to exhibit selectivity for 5-HT$_1$ receptors localized on large intracranial blood vessels, and to have little or no affinity for other neurotransmitter binding sites.[203,207,208] It is effective when administered either subcutaneously or orally during a migraine attack, and is more effective in relieving headache than other medications such as the ergot derivatives, which show less selectivity for 5-HT$_1$ receptors.[204,209]

A number of molecular subtypes of the pharmacologically defined 5-HT$_1$ receptor class have been described (see chapter 1, Table 1.1), and it is not yet clear which molecular subtype(s) mediates the actions of sumatriptan. Continuing investigations will undoubtedly delineate the mode of action of this therapeutic agent and may lead to the development of even more specific agonists.

## 3. ADENOSINE RECEPTORS

In addition to occupying a central position in cellular energy metabolism and energy transduction processes, the purine nucleoside adenosine functions as a neurotransmitter and neuromodulator. Through interaction with discrete receptor subtypes, adenosine has numerous and widespread effects in the central and peripheral nervous system, and mediates biological effects in many organ systems, including the cardiovascular, respiratory, gastrointestinal, renal and immune systems, as well as adipose tissue.[210] For example, adenosine is believed to play an important role in limitation of cell damage during periods of energy depletion. The release of adenosine from hyperactive or hypoxic cells results in a general depressant effect on cell activity and reduction in metabolic demand. Associated vasodilatory effects of adenosine on local vasculature, allowing for increased nutrient supply, provide a significant protective mechanism in the heart, while adenosine plays a related role in controlling damage from ischemia in the CNS through inhibition of release of excitotoxic neurotransmitters such as L-glutamate.[211]

The current major clinical use of adenosine is in the acute treatment and diagnosis of supraventricular arrhythmia, while the adenosine receptor antagonist theophylline is used in the treatment of asthma and other antagonists such as caffeine are widely used as stimulants. However, given the widespread distribution of adenosine receptors and the variety of biological effects mediated by this purine nucleoside, additional therapeutic applications for agonists have been suggested in a variety of metabolic, CNS and cardiovascular disorders, and for antagonists in treatment of renal disease, sleep apnea and cognitive disorders.[210] The development of specific therapeutic agents will rely on discrimination of biological effects mediated by discrete adenosine receptor subtypes and definition of sequence and structural determinants underlying receptor subtype selectivity. The molecular cloning of adenosine receptor subtypes provides a substantial contribution to further research along these lines.

## 4. Peptide Receptors

The superfamily of G protein-coupled receptors encompasses receptors for a large number of biologically active peptides and neuropeptides (see chapter 1, Table 1.1). It is becoming increasingly clear that many neuropeptides play important roles as primary and auxiliary messengers in neuronal signaling, and that the actions of the classical neurotransmitters may be modulated by both neurogenic and other peptides.[212] For example, important roles for peptides have been identified in complex behavioral and physiological processes such as appetite control and cardiovascular regulation. In addition, a number of peptides exhibit biological activities which may be of significance in inflammatory and immune responses, and in the pathophysiology of inflammatory disease.

### i. Eating disorders

A large number of neurochemical and neuroendocrine regulatory mechanisms involving peripheral organs and diverse regions of the brain mediate control of food intake, appetite and metabolism. Novel therapeutic strategies for behavioral disorders associated with food intake are incorporating

knowledge of the role of peptides in regulatory mechanisms. For example, among peptide effectors, CCK has been identified as mediating satiety in response to consumption of protein or fat, through activation of $CCK_A$ receptors in the pyloric region of the stomach. With demonstration that the administration of $CCK_A$ receptor antagonists leads to an increase in food intake in animals, much research activity has subsequently been directed towards the development of analogs specific for the $CCK_A$ receptor which are of anorexic potency.[213] Similarly, within the CNS, a role has been identified for bombesin, calcitonin, CCK, corticotropin-releasing factor (CRF), neurotensin, somatostatin, thyroptropin-releasing hormone (TRH) and vasoactive intestinal peptide (VIP) in suppression of feeding, and for galanin, neuropeptide Y (NPY), peptide YY (PYY) and the opioid peptides β-endorphin and dynorphin in increasing food intake.[213,214] While the complex interactions of these and other neurotransmitters in appetite regulation within the CNS remain to be defined, the development of specific analogs with appropriate receptor subtype selectivity represents an avenue of continuing investigation with potential therapeutic application in the treatment of eating disorders.

### ii. Cardiovascular regulation

The neuropeptides calcitonin gene-related peptide (CGRP), NPY, substance P, VIP and vasopressin have been shown to have potent vasoactive properties within the mammalian cardiovascular system. Other vasoactive peptides which interact with G protein-coupled receptors include angiotensin II, bradykinin and endothelin-1. Angiotensin II, endothelin-1, NPY and vasopressin are vasoconstrictors, while bradykinin, CGRP, substance P and VIP are vasodilators. The mechanisms by which the neuromodulatory and other peptides interact with classical neurotransmitter control of cardiovascular functioning is beginning to be appreciated with characterization of the distribution and functional capabilities of receptors mediating their effects and the development of specific agonists and antagonists.[215] Investigation of the roles of various mediators in cardiovascular regulation and pathophysiological conditions may provide the means for more directed therapeutic approaches in complex disorders such as hypertensive phenotypes.[216]

### iii. Immune responses and inflammatory disease

Recent work has identified a role for neuropeptides in modulation of inflammatory and immune responses and suggested their involvement in a number of inflammatory diseases. For example, the tachykinin peptide substance P has potent effects on proliferative and physiological responses of lymphocytes and macrophages and may be a factor in the pathophysiology of rheumatoid arthritis.[217] Substance P may also play a role in the pathogenesis of inflammatory bowel disease,[218] and neurogenic inflammation mediated by substance P and other neuropeptides may contribute to the inflammatory response in asthmatic airways.[219] A clear definition of the contribution of substance P and other neuropeptides in conditions characterized by tissue damage could represent an alternative means for therapeutic intervention, through the development of potent analogs which specifically target neuropeptide receptor function.

### 5. ANTISENSE THERAPIES

The classical means for modulation of receptor function makes use of specific agonist or antagonist binding for stimulation or inhibition of activity, respectively. An alternative approach for receptor down-regulation is possible with knowledge of the DNA sequence encoding an expressed receptor subtype. Specific interference at the level of synthesis of the receptor within a particular cell type may be achieved by the introduction of antisense oligodeoxynucleotide sequences. Antisense oligodeoxynucleotides have been used successfully to arrest mRNA translation in in vitro systems, with demonstration of the capability of cells to internalize exogenous oligodeoxynucleotides.[220-223] The applicability of this technology as a therapeutic approach in systemic disease has been well documented.[224]

Recent work has shown potential for administration of antisense oligodeoxynucleotides in studies of integrative function in the CNS. Intracerebroventricular (i.c.v.) infusion of antisense oligodeoxynucleotides to the angiotensin II $AT_1$, dopamine $D_2$ or NPY $Y_1$ receptor subtypes has been demonstrated to result in specific down-regulation of the targeted receptor subtype.[225-227] Behavioral correlates of receptor down-regulation have identified, respectively, the involvement of dopamine $D_2$ receptors in locomotor activation and NPY $Y_1$ receptors in anxiolytic mechanisms. Antisense inhibition of angiotensin II $AT_1$ receptor expression in the spontaneously hypertensive rat was correlated with a reduction in mean arterial blood pressure and normalization of the hypertensive phenotype. It is possible that antisense strategies may see general clinical application in circumstances where reduced expression of a particular receptor subtype is of therapeutic benefit.

### SUMMARY

Naturally occurring mutations have now been described in a number of G protein-coupled receptors. These include molecular defects in rhodopsin and the visual pigment proteins which are responsible for visual disturbances such as those encountered in retinitis pigmentosa and other disorders of vision, mutations in the vasopressin $V_2$ receptor which give rise to nephrogenic diabetes insipidus and mutations in the MSH receptor which underlie phenotypic differences in skin pigmentation. A single mutation in the LH/CG receptor leads to constitutive receptor activation and is manifested as the pathological condition familial male precocious puberty, while mutations leading to defective receptor function in the ACTH receptor underlie familial glucocorticoid deficiency and in the $Ca^{2+}$-sensing receptor result in familial hypocalciuric hypercalcemia and neonatal severe hyperparathyroidism.

It has been known for some time that G protein-coupled receptors and the signaling pathways they activate play a role in cell growth and differentiation, and may be responsible for malignant transformation. For example, considerable insight has been gained into the involvement of G protein-coupled receptors in fertilization and gamete activation. Similarly, the crucial role of GHRH and TSH receptor in normal pituitary and thyroid development, respectively, is highlighted by developmental abnormalities that are associated with aberrant expression or dysfunction of these receptors. Particular functional properties of receptors such as the muscarinic acetylcholine and serotonin receptor subtypes allow them to act as agonist-dependent oncogenes, while constitutively active mutants of the TSH receptor have recently been shown to be associated with hyperfunctioning thyroid adenoma. The identification of a range of G protein-coupled receptors on malignant cells which may play a role in mitogenesis raises possibilities for therapeutic intervention and diagnostic applications. In addition, there are a number of viral receptors which interact with mammalian ligands and may play a role in proliferative processes. G protein-coupled receptors are also the target of autoimmune responses, as demonstrated by characterization of autoantibodies to the TSH receptor which are responsible for the pathology of Graves' disease and idiopathic myxedema.

Genetic linkage analyses in different disorders have been performed with molecular probes for a number of receptors. These have included neurotransmitter receptors which are of potential significance in neurological and neuropsychiatric disorders such as bipolar affective disorder, schizophrenia, Gilles de la Tourette syndrome and alcoholism. Given an expanding appreciation of receptor subtype diversity, the efficacy of therapeutic agents such as clozapine, which is used in the treatment of schizophrenia, and sumatriptan,

which is effective in the treatment of acute migraine headache attacks, may now be correlated with their specificity for particular receptor subtypes. Novel therapeutic strategies will undoubtedly incorporate an increasing awareness of the involvement of a variety of mediators in physiological and pathophysiological processes, with the description of discrete molecular subtypes of receptors providing the basis for development of subtype-specific pharmacological agents.

## REFERENCES

1. Lindsay S, Inglehearn CF, Curtis A et al. Molecular genetics of inherited retinal degenerations. Curr Opin Genet Dev 1992; 2:459-466.
2. Shokravi MT, Dryja TP. Retinitis pigmentosa and the rhodopsin gene. Int Ophthalmol Clin 1993; 33:219-228.
3. McWilliam P, Farrar GJ, Kenna P et al. Autosomal dominant retinitis pigmentosa (ADRP): localization of an ADRP gene to the long arm of chromosome 3. Genomics 1989; 5:619-622.
4. Macke JP, Davenport CM, Jacobson SG et al. Identification of novel rhodopsin mutations responsible for retinitis pigmentosa: implications for the structure and function of rhodopsin. Am J Hum Genet 1993; 53:80-89.
5. Sullivan JM, Scott KM, Falls HF et al. A novel rhodopsin mutation at the retinal binding site (Lys-296-Met) in ADRP. Invest Ophthalmol Vis Sci 1993; 34:1149.
6. Rao VR, Cohen GB, Oprian DD. Rhodopsin mutation G90D and a molecular mechanism for congenital night blindness. Nature 1994; 367:639-642.
7. Rosenfeld PJ, Cowley GS, McGee TL et al. A *Null* mutation in the rhodopsin gene causes rod photoreceptor dysfunction and autosomal recessive retinitis pigmentosa. Nature Genet 1992; 1:209-213.
8. Sung C-H, Schneider BG, Agarwal N et al. Functional heterogeneity of mutant rhodopsins responsible for autosomal dominant retinitis pigmentosa. Proc Natl Acad Sci USA 1991; 88:8840-8844.
9. Robinson PR, Cohen GB, Zhukovsky EA

et al. Constitutively active mutants of rhodopsin. Neuron 1992; 9:719-725.

10. Govardhan CP, Oprian DD. Active site-directed inactivation of constitutively active mutants of rhodopsin. J Biol Chem 1994; 269:6542-6547.

11. Olsson JE, Gordon JW, Pawlyk BS et al. Transgenic mice with a rhodopsin mutation (Pro23His): a mouse model of autosomal dominant retinitis pigmentosa. Neuron 1992; 9:815-830.

12. Dryja TP, Berson EL, Rao VR et al. Heterozygous missense mutation in the rhodopsin gene as a cause of congenital stationary night blindness. Nature Genet 1993; 4:280-283.

13. Nathans J, Davenport CM, Maumenee IH et al. Molecular genetics of human blue cone monochromacy. Science 1989; 245:831-838.

14. Weitz CJ, Miyake Y, Shinzato K et al. Human tritanopia associated with two amino acid substitutions in the blue-sensitive opsin. Am J Hum Genet 1992; 50:498-507.

15. Nakano KK. Familial nephrogenic diabetes insipidus. Hawaii Med J 1969; 28:205-208.

16. Langley JM, Balfe JW, Selander T et al. Autosomal recessive inheritance of vasopressin-resistant diabetes insipidus. Am J Med Genet 1991; 38:90-94.

17. Naik DV, Valtin H. Hereditary vasopressin-resistant urinary concentrating defects in mice. Am J Physiol 1969; 217:1183-1190.

18. Knoers N, van der Heyden H, van Oost BA et al. Nephrogenic diabetes insipidus: close linkage with markers from the distal long arm of the human X chromosome. Hum Genet 1988; 80:31-38.

19. Jans DA, van Oost BA, Ropers HH et al. Derivatives of somatic cell hybrids which carry the human gene locus for nephrogenic diabetes insipidus (ND) express functional vasopressin renal $V_2$-type receptors. J Biol Chem 1990; 265:15379-15382.

20. Birnbaumer M, Seibold A, Gilbert S et al. Molecular cloning of the receptor for human antidiuretic hormone. Nature 1992; 357:333-335.

21. Lolait SJ, O'Carroll A-M, McBride OW et al. Cloning and characterization of a vasopressin V2 receptor and possible link to nephrogenic diabetes insipidus. Nature 1992; 357:336-339.

22. van den Ouweland AM, Knoop MT, Knoers VV et al. Colocalization of the gene for nephrogenic diabetes insipidus (DIR) and the vasopressin type 2 receptor gene (AVPR2) in the Xq28 region. Genomics 1992; 13:1350-1352.

23. Rosenthal W, Seibold A, Antaramian A et al. Molecular identification of the gene responsible for congenital nephrogenic diabetes insipidus. Nature 1992; 359:233-235.

24. van den Ouweland AMW, Dreesen JCFM, Verdijk M et al. Mutations in the vasopressin type 2 receptor gene (AVPR2) associated with nephrogenic diabetes insipidus. Nature Genet 1992; 2:99-102.

25. Pan Y, Metzenberg A, Das S et al. Mutations in the V2 vasopressin receptor gene are associated with X-linked nephrogenic diabetes insipidus. Nature Genet 1992; 2:103-106.

26. Holtzman E, Harris HW, Kolakowski LF et al. A molecular defect in the vasopressin $V_2$ receptor gene causing nephrogenic diabetes insipidus. New Eng J Med 1993; 328:1534-1537.

27. Merendino JJ, Spiegel AM, Crawford JD et al. A mutation in the vasopressin V2-receptor gene in a kindred with X-linked nephrogenic diabetes insipidus. New Engl J Med 1993; 328:1538-1541.

28. Rosenthal W, Antaramian A, Gilbert S et al. Nephrogenic diabetes insipidus. A V2 vasopressin receptor unable to stimulate adenylyl cyclase. J Biol Chem 1993; 268:13030-13033.

29. Tsukaguchi H, Matsubara H, Aritaki S et al. Two novel mutations in the vasopressin V2 receptor gene in unrelated Japanese kindreds with nephrogenic diabetes insipidus. Biochem Biophys Res Commun 1993; 197:1000-1010.

30. Sharif M, Hanley MR. Peptide receptors: stepping up the pressure. Nature 1992; 357:279-280.

31. Robbins LS, Nadeau JH, Johnson KR et al. Pigmentation phenotypes of variant extension locus alleles result from point mutations that alter MSH receptor function. Cell 1993; 72:827-834.

32. Shenker A, Laue L, Kosugi S et al. A constitutively activating mutation of the luteinizing hormone receptor in familial precocious puberty. Nature 1993; 365:652-654.

33. Clark AJL, McLoughlin L, Grossman A. Familial glucocorticoid deficiency associated with point mutation in the adrenocorticotropin receptor. Lancet 1993; 341:461-462.

34. Weber A, Kapas S, Hinson J et al. Functional characterization of the cloned human ACTH receptor: impaired responsiveness of a mutant receptor in familial glucocorticoid deficiency. Biochem Biophys Res Commun 1993; 197:172-178.

35. Chhajlani V, Muceniece R, Wikberg JES. Molecular cloning of a novel human melanocortin receptor. Biochem Biophys Res Commun 1993; 195:866-873.

36. Gantz I, Miwa H, Konda Y et al. Molecular cloning, expression, and gene localization of a fourth melanocortin receptor. J Biol Chem 1993; 268:15174-15179.

37. Tsigos C, Arai K, Hung W et al. Hereditary isolated glucocorticoid deficiency is associated with abnormalities of the adrenocorticotropin receptor gene. J Clin Invest 1993; 92:2458-2461.

38. Clapham DE. Mutations in G protein-linked receptors: novel insights on disease. Cell 1993; 75:1237-1239.

39. Pollak MR, Brown EM, Chou Y-HW et al. Mutations in the human $Ca^{2+}$-sensing receptor gene cause familial hypocalciuric hypercalcemia and neonatal severe hyperparathyroidism. Cell 1993; 75:1297-1303.

40. Brown EM, Gamba G, Riccardi D et al. Cloning and characterization of an extracellular $Ca^{2+}$-sensing receptor from bovine parathyroid. Nature 1993; 366:575-580.

41. Kopf GS, Endo Y, Mattei P et al. Egg-induced modifications of the murine *zona pellucida*. In: Nuccitelli R, Cherr GN, Clark WH, ed. Mechanisms of Egg Activation. New York: Plenum, 1989:249-272.

42. Kopf GS, Gerton GL. The mammalian sperm acrosome and the acrosome reaction. In: Wassarman PM, ed. The Biology and Chemistry of Fertilization. Florida: CRC Press, Boca Raton, 1991:153-203.

43. Blobel CP, Wolfsberg TG, Turck CW et al. A potential fusion peptide and an integrin domain in a protein active in sperm-egg fusion. Nature 1992; 356:248-252.

44. Cran DG, Moor RM, Irvine RF. Initiation of the cortical reaction in hamster and sheep oocytes in response to inositol triphosphates. J Cell Sci 1988; 91:139-144.

45. Miyasaki S. Inositol 1,4,5-triphosphate-induced calcium release and guanine nucleotide-binding protein-mediated periodic calcium rises in golden hamster eggs. J Cell Biol 1988; 106:345-353.

46. Endo Y, Schultz RM, Kopf GS. Effects of phorbol esters and a diacylglycerol on mouse eggs: inhibition of fertilization and modifications of the *zona pellucida*. Dev Biol 1987; 119:199-209.

47. Kurasawa S, Schultz RM, Kopf GS. Egg-induced modifications of the *zona pellucida* of the mouse egg: effects of microinjected 1,4,5-trisphosphate. Dev Biol 1989; 133:295-304.

48. Moore GD, Kopf GS, Schultz RM. Complete mouse egg activation in the absence of sperm by stimulation of an exogenous G protein-coupled receptor. Dev Biol 1993; 159:669-678.

49. Ganong WF. Review of Medical Physiology. 14th Edition. New Jersey: Prentice-Hall, 1989.

50. Billestrup N, Swanson LW, Vale W. Growth hormone-releasing factor stimulates proliferation of somatotrophs in vitro. Proc Natl Acad Sci USA 1986; 83:6854-6857.

51. Mayo KE, Hammer RE, Swanson LW et al. Dramatic pituitary hyperplasia in transgenic mice expressing a human growth hormone-releasing factor gene. Mol Endocrinol 1988; 2:606-612.

52. Burton FH, Hasel KW, Bloom FE et al. Pituitary hyperplasia and gigantism in mice caused by a cholera toxin transgene. Nature 1991; 350:74-77.

53. Struthers RS, Vale WW, Arias C et al. Somatotroph hypoplasia and dwarfism in transgenic mice expressing a nonphosphorylatable CREB mutant. Nature 1991; 350:622-624.

54. Ingraham HA, R.P. C, Mangalam HJ et al. A tissue-specific transcription factor containing a homeodomain specifies a pituitary phenotype. Cell 1988; 55:519-529.

55. Simmons DM, Voss JW, Ingraham HA et al. Pituitary cell phenotypes involve cell-specific Pit-1 mRNA translation and synergistic interactions with other classes of transcription factors. Genes Dev 1990; 4:695-711.

56. Roux M, Bartke A, Dumont F et al. Immunohistological study of the anterior pituitary gland - pars distalis and pars intermedia - in dwarf mice. Cell Tissue Res 1982; 223:415-420.

57. Wilson DB, Wyatt DP. Immunocytochemistry of TSH cells during development of the dwarf mutant mouse. Anat Embryol 1986; 174:277-282.

58. Yashiro T, Arai M, Miyashita E et al. Fine-structural and immunohistochemical study of anterior pituitary cells of Snell dwarf mice. Cell Tissue Res 1988; 251:249-255.

59. Li S, Crenshaw EB, Rawson EJ et al. Dwarf locus mutants lacking three pituitary cell types result from mutations in the POU-domain gene pit-1. Nature 1990; 247:528-533.

60. Lin C, Lin S-C, Chang C-P et al. Pit-1-dependent expression of the receptor for growth hormone releasing factor mediates cell growth. Nature 1992; 360:765-768.

61. Mayo KE. Molecular cloning and expression of a pituitary-specific receptor for growth hormone-releasing hormone. Mol Endocrinol 1992; 6:1734-1744.

62. Eicher EM, Beamer WG. Inherited ateliotic dwarfism in mice. Characteristics of the mutation, little, on chromosome 6. J Hered 1976; 67:87-91.

63. Jansson JO, Downs TR, Beamer WG et al. Receptor-associated resistance to growth hormone-releasing factor in dwarf "little" mice. Science 1986; 232:511-512.

64. Godfrey P, Rahal JO, Beamer WG et al. GHRH receptor of *little* mice contains a missense mutation in the extracellular domain that disrupts receptor function. Nature Genet 1993; 4:227-231.

65. Lin S-C, Lin CR, Gukovsky I et al. Molecular basis of the little mouse phenotype and implications for cell type-specific growth. Nature 1993; 364:208-213.

66. Ishihara T, Nakamura S, Kaziro Y et al. Molecular cloning and expression of a cDNA encoding the secretin receptor. 1991; 10:1635-1641.

67. Stein SA, Oates EL, Hall CR et al. Identification of a point mutation in the thyrotropin receptor of the *hyt/hyt* mouse. Mol Endocrinol 1994; 8:129-138.

68. Young D, Waitches G, Birchmeier C et al. Isolation and characterization of a new cellular oncogene encoding a protein with multiple potential transmembrane domains. Cell 1986; 45:711-719.

69. van't Veer LJ, van den Berg-Bakker LAM, Hermans RPMG et al. High frequency of *mas* oncogene activation detected in the NIH3T3 tumorigenicity assay. Oncogene Res 1988; 3:247-254.

70. Jackson TR, Blair LAC, Marshall J et al. The *mas* oncogene encodes an angiotensin receptor. Nature 1988; 335:437-440.

71. Ambroz C, Clark AJ, Catt KJ. The *mas* oncogene enhances angiotensin-induced $[Ca^{2+}]_i$ responses in cells with pre-existing angiotensin II receptors. Biochim Biophys Acta 1991; 1133:107-111.

72. Gutkind JS, Novotny EA, Brann MR et al. Muscarinic acetylcholine receptor subtypes as agonist-dependent oncogenes. Proc Natl Acad Sci USA 1991; 88:4703-4707.

73. Watson S, Girdlestone D. Receptor & Ion Channel Nomenclature Supplement. Trends Pharmacol Sci 1994; 15 Suppl:1-51.

74. Julius D, Livelli TJ, Jessell TM et al. Ectopic expression of the serotonin 1c receptor and the triggering of malignant transformation. Science 1989; 244:1057-1062.

75. Abdel-Baset H, Bozovic V, Szyf M et al. Conditional transformation mediated via a pertussis toxin-sensitive receptor signalling pathway. Mol Endocrinol 1992; 6:730-740.

76. Ives HE. GTP binding proteins and growth factor signal transduction. Cell Signal 1991; 3:491-499.

77. Zachary I, Gil J, Lehmann W et al. Bombesin, vasopressin, and endothelin rapidly stimulate tyrosine phosphorylation in intact Swiss 3T3 cells. Proc Natl Acad Sci USA 1991; 88:4577-4581.

78. Simonson MS, Herman WH. Protein kinase C and protein tyrosine kinase activity contribute to mitogenic signalling by endothelin-1. Cross-talk between G protein-

coupled receptors and pp60$^{c\text{-}src}$. J Biol Chem 1993; 268:9347-9357.

79. Gutkind JS, Robbins KC. Activation of transforming G protein-coupled receptors induces rapid tyrosine phosphorylation of cellular proteins, including p125FAK and the p130 v-src substrate. Biochem Biophys Res Commun 1992; 188:155-161.

80. Stephens EV, Kalinec G, Brann MR et al. Transforming G protein-coupled receptors transduce potent mitogenic signals in NIH 3T3 cells independent on cAMP inhibition or conventional protein kinase C. Oncogene 1993; 8:19-26.

81. Berra E, Diaz-Meco MT, Dominguez I et al. Protein kinase C ζ isoform is critical for mitogenic signal transduction. Cell 1993; 74:555-563.

82. Allen LF, Lefkowitz RJ, Caron MG et al. G-protein-coupled receptor genes as proto-oncogenes: constitutively activating mutation of the $\alpha_{1B}$-adrenergic receptor enhances mitogenesis and tumorigenicity. Proc Natl Acad Sci USA 1991; 88:11354-11358.

83. Parma J, Duprez L, Van Sande J et al. Somatic mutations in the thyrotropin receptor gene cause hyperfunctioning thyroid adenomas. Nature 1993; 365:649-651.

84. Schüller HM. Receptor-mediated mitogenic signals and lung cancer. Cancer Cells 1991; 3:496-503.

85. Sethi T, Langdon S, Smyth J et al. Growth of small cell lung cancer cells: stimulation by multiple neuropeptides and inhibition by broad spectrum antagonists *in vitro* and *in vivo*. Cancer Res (Suppl) 1992; 52:2737s-2742s.

86. Moody TW, Cuttitta F. Growth factor and peptide receptors in small cell lung cancer. Life Sci 1993; 52:1161-1173.

87. Sethi T, Rozengurt E. Multiple neuropeptides stimulate clonal growth of small cell lung cancer: effects of bradykinin, vasopressin, cholecystokinin, galanin, and neurotensin. Cancer Res 1991; 51:3621-3623.

88. Cuttitta F, Carney DN, Mulshine J et al. Bombesin-like peptides can function as autocrine growth factors in human small-cell lung cancer. Nature 1985; 316:823-826.

89. Bepler G, Zeymer U, Mahmoud S et al. Substance P analogues function as bombesin

receptor antagonists and inhibit small cell lung cancer clonal growth. Peptides 1989; 9:1367-1372.

90. Langdon S, Sethi T, Ritchie A et al. Broad spectrum neuropeptide antagonists inhibit the growth of small cell lung cancer *in vivo*. Cancer Res 1992; 52:4554-4557.

91. Thomas F, Arvelo F, Antoine E et al. Antitumoral activity of bombesin analogues on small cell lung cancer xenografts: relationship with bombesin receptor expression. Cancer Res 1992; 52:4872-4877.

92. Zhou W, Povoski SP, Bell RH. Over-expression of messenger RNA for cholecystokinin-A receptor and novel expression of messenger RNA for gastrin (cholecystokinin-B) receptor in azaserine-induced rat pancreatic carcinoma. Carcinogenesis 1993; 14:2189-2192.

93. Reubi JC, Weber B, Khosla S et al. In vitro and in vivo detection of somatostatin receptors in pheochromocytomas and paragangliomas. J Clin Endocrinol Metab 1992; 74:1082-1089.

94. van Eijck CH, Krenning EP, Boostma A et al. Somatostatin-receptor scintigraphy in primary breast cancer. Lancet 1994; 343:640-643.

95. Chee MS, Satchwell C, Preddie E et al. Human cytomegalovirus encodes three G protein-coupled receptors. Nature 1990; 344:774-777.

96. Neote K, DiGregorio D, Mak JY et al. Molecular cloning, functional expression, and signaling characteristics of a C-C chemokine receptor. Cell 1993; 72:415-425.

97. Ahuja SA, Murphy PM. Molecular piracy of mammalian interleukin-8 receptor type B by herpesvirus saimiri. J Biol Chem 1993; 268:20691-20694.

98. Nicholas J, Cameron KR, Honess RW. Herpesvirus saimiri encodes homologues of G protein-coupled receptors and cyclins. Nature 1992; 355:362-365.

99. Anderson DM, Patwell JM, Plaut K et al. Dorland's Illustrated Medical Dictionary. 27th Edition. Philadelphia: WB Saunders, 1988.

100. Tzartos SJ, Barkas T, Cung MT et al. The main immunogenic region of the acetylcholine receptor. Structure and role in

myasthenia gravis. Autoimmunity 1991; 8:259-270.

101. Protti MP, Manfredi AA, Horton RM et al. Myasthenia gravis: recognition of a human autoantigen at the molecular level. Immunol Today 1993; 14:363-368.

102. Graus YM, De Baets MH. Myasthenia gravis: an autoimmune response against the acetylcholine receptor. Immunol Res 1993; 12:78-100.

103. Amino N. Receptors in disease: an overview. Clin Biochem 1990; 23:31-36.

104. Mooij P, Drexhage HA. Interactions between the immune system and the thyroid. Regulatory networks in health and disease. Thyroidology 1992; 4:45-48.

105. Gupta MK. Thyrotropin receptor antibodies: advances and importance of detection techniques in thyroid disease. Clin Biochem 1992; 25:193-199.

106. Feliciano DV. Everything you wanted to know about Graves' disease. Am J Surg 1992; 164:404-411.

107. Kohn LD, Kosugi S, Ban T et al. Molecular basis for the autoreactivity against thyroid stimulating hormone receptor. Int Rev Immunol 1992; 9:135-165.

108. Martin JB. Molecular genetics: applications to the clinical neurosciences. Science 1987; 238:765-772.

109. Collins FS. Cystic fibrosis: molecular biology and therapeutic implications. Science 1992; 256:774-779.

110. Richards RI, Sutherland GR. Fragile X syndrome: the molecular picture comes into focus. Trends Genet 1992; 8:249-255.

111. Ross CA, McInnis MG, Margolis RL et al. Genes with triplet repeats: candidate mediators of neuropsychiatric disorders. Trends Neurosci 1993; 16:254-260.

112. Gershon E. Manic-depressive Illness. In: Goodwin FK, Jamison KR, ed. Genetics. New York: Oxford University Press, 1990:373-401.

113. Mendlewicz J, Sevy S, Mandelbaum K. Molecular genetics in affective illness. Life Sci 1993; 52:231-242.

114. Mitchell P, Mackinnon A, Waters B. The genetics of bipolar disorder. Aust NZ J Psychiatry 1993; 27:560-580.

115. Egeland JA, Gerhard DS, Pauls DL et al.

Bipolar affective disorders linked to DNA markers on chromosome 11. Nature 1987; 325:783-787.

116. Hodgkinson S, Sherrington R, Gurling H et al. Molecular genetic evidence for heterogeneity in manic depression. Nature 1987; 325:805-806.

117. Detera-Wadleigh SD, Berrettini W, Goldin LR et al. Close linkage of c-Harvey-ras 1 and the insulin gene to affective disorder is ruled out in three North American pedigrees. Nature 1987; 325:806-808.

118. Gill M, McKeon P, Humphries P. Linkage analysis of manic depression in an Irish family using H-ras 1 and INS DNA markers. J Med Genet 1988; 25:634-635.

119. Kelsoe JR, Ginns EI, Egeland JA et al. Re-evaluation of the linkage relationship between chromosome 11p loci and the gene for bipolar affective disorder in the old Order Amish. Nature 1989; 342:238-243.

120. Wesner RB, Scheftner W, Palmer PJ et al. The effect of co-morbidity and penetrance estimation on the outcome of linkage analysis in a bipolar family. Biol Psychiatry 1990; 27:241-244.

121. Mendlewicz J, Leboyer M, DeBruyn A et al. Absence of linkage between chromosome 11p15 markers and manic-depressive illness in a Belgian pedigree. Am J Psychiatry 1991; 148:1683-1687.

122. Mitchell P, Waters B, Morrison N et al. Close linkage of bipolar disorder to chromosome 11 markers is excluded in two large Australian pedigrees. J Affect Disord 1991; 21:23-32.

123. Pakstis AJ, Kidd JR, Castiglione CM et al. Status of the search for a major genetic locus for affective disorder in the Old Order Amish. Hum Genet 1991; 87:475-483.

124. Ginns EI, Egeland JA, Allen CR et al. Update on the search for DNA markers linked to manic-depresive illness in the Old Order Amish. J Psychiat Res 1992; 26:305-308.

125. Berrettini WH, Detera-Wadleigh SD, Goldin LR et al. Genomic screening for genes predisposing to bipolar disease. Psychiatric Genet 1991; 2:191-208.

126. Detera-Wadleigh SD, Berrettini WH, Goldin LR et al. A systematic search for a

bipolar predisposing locus on chromosome 5. Neuropsychopharmacol 1992; 6:219-229.

127. Gejman PV, Martinez M, Cao Q et al. Linkage analysis of fifty-seven microsatellite loci to bipolar disorder. Neuropsychopharmacol 1993; 9:31-40.

128. Curtis D, Sherrington R, Brett P et al. Genetic linkage analysis of manic depression in Iceland. J R Soc Med 1993; 86:506-510.

129. Coon H, Jensen S, Hoff M et al. A genome-wide search for genes predisposing to manic depression, assuming autosomal dominant inheritance. Am J Hum Genet 1993; 52:1234-1249.

130. Byerley W, Leppert M, O'Connell P et al. $D_2$ dopamine receptor gene not linked to manic-depression in three families. Psychiatric Genet 1990; 1:55-62.

131. Holmes D, Brynjolfsson J, Brett P et al. No evidence for a susceptibility locus predisposing to manic depression in the region of the dopamine (D2) receptor gene. Br J Psychiatry 1991; 158:635-641.

132. Jensen S, Plaetke R, Holik J et al. Linkage analysis of the D1 dopamine receptor gene and manic depression in six families. Hum Hered 1992; 42:269-275.

133. Mitchell P, Selbie L, Waters B et al. Exclusion of close linkage of bipolar disorder to dopamine $D_1$ and $D_2$ receptor gene markers. J Affect Disord 1992; 25:1-12.

134. Nöthen MM, Erdmann J, Korner J et al. Lack of association between dopamine $D_1$ and $D_2$ receptor genes and bipolar affective disorder. Am J Psychiatry 1992; 149:199-201.

135. Mitchell P, Waters B, Vivero C et al. Exclusion of close linkage of bipolar disorder to the dopamine $D_3$ receptor gene in nine Australian pedigrees. J Affect Disord 1993; 27:213-224.

136. Shaikh S, Ball D, Craddock N et al. The dopamine D3 receptor gene: no association with bipolar affective disorder. J Med Genet 1993; 30:308-309.

137. De bruyn A, Mendelbaum K, Sandkuijl LA et al. Nonlinkage of bipolar illness to tyrosine hydroxylase, tyrosinase, $D_2$ and $D_4$ dopamine receptor genes on chromosome 11. Am J Psychiatry 1994; 151:102-106.

138. Curtis D, Brynjolfsson J, Petursson H et al. Segregation and linkage analysis in five manic depression pedigrees excludes the 5HT1a receptor gene (*HTR1A*). Ann Hum Genet 1993; 57:27-39.

139. Le F, Mitchell P, Vivero C et al. Exclusion of close linkage of bipolar disorder to the $G_s$ α-subunit in nine Australian pedigrees. J Affect Disord 1993; *in press*.

140. Carpenter WT, Buchanan RW. Schizophrenia. New Engl J Med 1994; 330:681-690.

141. Kendler KS. Overview: a current perspective on twin studies of schizophrenia. Am J Psychiatry 1983; 140:1413-1425.

142. Rao DC, Morton NE, Gottesman II et al. Path analysis of qualitative data on pairs of relatives: application to schizophrenia. Hum Hered 1981; 31:325-333.

143. McGue M, Gottesman II, Rao DC. The transmission of schizophrenia under a multifactorial threshold model. Am J Hum Genet 1983; 35:1161-1178.

144. McGue M, Gottesman II, Rao DC. Resolving genetic models for the transmission of schizophrenia. Genet Epidemiol 1985; 2:99-110.

145. Mowry BJ, Levinson DF. Genetic linkage and schizophrenia: methods, recent findings and future directions. Aust NZ J Psychiatry 1993; 27:200-218.

146. Di Chiara G, Imperato A. Drugs abused by humans preferentially increase synaptic dopamine concentrations in the mesolimbic system of freely moving rats. Proc Natl Acad Sci USA 1988; 85:5274-5278.

147. Seeman P. Schizophrenia as a brain disease. The dopamine receptor story. Arch Neurol 1993; 50:1093-1095.

148. Seeman P, Guan H-C, Van Tol HHM. Dopamine D4 receptors elevated in schizophrenia. Nature 1993; 365:441-445.

149. Seeman P. Dopamine receptor sequences. Therapeutic levels of neuroleptics occupy D2 receptors, clozapine occupies D4. Neuropsychopharmacol 1992; 7:261-284.

150. Goldstein M, Deutch AY. Dopaminergic mechanisms in the pathogenesis of schizophrenia. FASEB J 1992; 6:2413-2421.

151. Jensen S, Plaetke R, Hoff M et al. Linkage analysis of schizophrenia: the D1 dopamine receptor gene and several flanking DNA

markers. Hum Hered 1993; 43:58-62.

152. Su Y, Burke J, O'Neill FA et al. Exclusion of linkage between schizophrenia and the D2 dopamine receptor gene region of chromosome 11q in 112 Irish multiplex families. Arch Gen Psychiatry 1993; 50:205-211.

153. Moises HW, Gelernter J, Giuffra LA et al. No linkage between $D_2$ dopamine receptor gene region and schizophrenia. Arch Gen Psychiatry 1991; 48:643-647.

154. Wiese C, Lannfelt L, Kristbjarnarson H et al. No evidence of linkage between schizophrenia and D3 dopamine receptor gene locus in Icelandic pedigrees. Psychiatry Res 1993; 46:69-78.

155. Yang L, Li T, Wiese C et al. No association between schizophrenia and homozygosity at the D3 dopamine receptor gene. Am J Med Genet 1993; 48:83-86.

156. Nanko S, Sasaki T, Fukuda R et al. A study of the association between schizophrenia and the dopamine $D_3$ receptor gene. Hum Genet 1993; 92:336-338.

157. Kennedy JL, Van Tol HM, Petronis A et al. Dopamine $D_4$ receptor variants and schizophrenia. Am J Hum Genet (Suppl) 1992; 51:A365.

158. Coon H, Byerley W, Holik J et al. Linkage analysis of schizophrenia with five different dopamine receptor genes in nine pedigrees. Am J Hum Genet 1993; 52:327-334.

159. Hallmayer J, Kennedy JL, Wetterberg L et al. Exclusion of linkage between the serotonin2 receptor and schizophrenia in a large Swedish kindred. Arch Gen Psychiatry 1992; 49:216-219.

160. O'Hara K, Ulpian C, Seeman P et al. Schizophrenia: dopamine $D_1$ receptor sequence is normal, but has DNA polymorphisms. Neuropsychopharmacol 1993; 8:131-135.

161. Sarkar G, Kapelner S, Grandy DK et al. Direct sequencing of the dopamine D2 receptor (DRD2) in schizophrenics reveals three polymorphisms but no structural change in the receptor. Genomics 1991; 11:8-14.

162. Seeman P, Ohara K, Ulpian C et al. Schizophrenia: normal sequence in the dopamine $D_2$ receptor region that couples to G-proteins. DNA polymorphisms in $D_2$.

Neuropsychopharmacol 1993; 8:137-142.

163. Itokawa M, Arinami T, Futamura N et al. A structural polymorphism of human dopamine D2 receptor. D2(Ser[311]-Cys). Biochem Biophys Res Commun 1993; 196:1369-1375.

164. Catalano M, Nobile M, Novelli E et al. Distribution of a novel mutation in the first exon of the human dopamine $D_4$ receptor gene in psychotic patients. Biol Psychiatry 1993; 34:459-464.

165. Shaikh S, Collier D, Kerwin RW et al. Dopamine D4 receptor subtypes and response to clozapine. Lancet 1993; 341: 116-117.

166. Nanko S, Hattori M, Ikeda K et al. Dopamine D4 receptor polymorphism and schizophrenia. Lancet 1993; 341:689-690.

167. Sedvall G. The current status of PET scanning with respect to schizophrenia. Neuropsychopharmacol 1992; 7:41-54.

168. Seeman P. Elevated $D_2$ in schizophrenia: role of endogenous dopamine and cerebellum. Commentary on "The current status of PET scanning with respect to schizophrenia". Neuropsychopharmacol 1992; 7:55-57.

169. Murphy F, Fitzgerald G. Gilles de la Tourette's syndrome: a case study. Axone 1992; 14:41-45.

170. Robertson MM, Channon S, Baker J et al. The psychopathology of Gilles de la Tourette's syndrome. A controlled study. Br J Psychiatry 1993; 162:114-117.

171. Sandor P. Gilles de la Tourette syndrome: a neuropsychiatric disorder. J Psychosomatic Res 1993; 37:211-226.

172. van de Wetering BJM, Heutnik P. The genetics of the Gilles de la Tourette syndrome: a review. J Lab Clin Med 1993; 11:638-645.

173. Regeur L. Clinical evaluation and pharmacological treatment of Gilles de la Tourette's syndrome and other hyperkinesias. Acta Neurol Scand 1992; 137:48-50.

174. Gelernter J, Kennedy JL, Grandy DK et al. Exclusion of close linkage of Tourette's syndrome to D1 dopamine receptor. Am J Psychiatry 1993; 150:449-453.

175. Devor EJ, Grandky DK, Civelli O et al. Genetic linkage is excluded for the $D_2$-dopamine receptor $\lambda HD_2G_1$ and flanking

loci on chromosome 11q22-q23 in Tourette syndrome. Hum Hered 1990; 40:105-108.

176. Gelernter J, Pakstis AJ, Pauls DL et al. Gilles de la Tourette syndrome is not linked to D$_2$-dopamine receptor. Arch Gen Psychiat 1990; 47:1073-1077.

177. Brett P, Robertson M, Gurling H et al. Failure to find linkage and increased homozygosity for the dopamine D3 receptor gene in Tourette's syndrome. Lancet 1993; 341:1225.

178. Nestler EJ. Molecular mechanisms of drug addiction. J Neurosci 1992; 12:2439-2450.

179. Loh HH, Smith AP. Molecular characterization of opioid receptors. Annu Rev Pharmacol Toxicol 1990; 30:123-147.

180. Clouet D, Asghar K, Brown R, ed. Mechanisms of cocaine abuse and toxicity. NIDA Research Monograph. Vol 88. Rockville: National Institute on Drug Abuse, 1988.

181. Liebman JM, Cooper SJ, eds. The neuropharmacological basis of reward. New York: Oxford University Press, 1989.

182. Peris J, Boyson SJ, Cass WA et al. Persistence of neurochemical changes in dopamine systems after repeated cocaine administration. J Pharmacol Exp Ther 1990; 253:38-44.

183. Ritz MC, Lamb RJ, Goldberg SR et al. Cocaine receptors on dopamine transporters are related to self-administration of cocaine. Science 1987; 237:1219-1223.

184. Persico AM, Vandenbergh DJ, Smith SS et al. Dopamine transporter gene polymorphisms are not associated with polysubstance abuse. Biol Psychiatry 1993; 34:265-267.

185. Blum K, Noble EP, Sheridan PJ et al. Allelic association of human dopamine D$_2$ receptor gene in alcoholism. J Am Med Assoc 1990; 263:2055-2060.

186. Noble EP. The D2 dopamine receptor gene: a review of association studies in alcoholism. Behav Genet 1993; 23:119-129.

187. Bolos AM, Dean M, Lucas-Derse S et al. Population and pedigree studies reveal a lack of association between the dopamine D2 receptor gene and alcoholism. J Am Med Assoc 1990; 264:3156-3160.

188. Gelernter J, O'Malley S, Risch N et al. No association between an allele at the D2 dopamine receptor gene (DRD2) and alcoholism. J Am Med Assoc 1991; 266:1801-1808.

189. Turner E, Ewing J, Shilling P et al. Lack of association between and RFLP near the D2 dopamine receptor gene and severe alcoholism. Biol Psychiatry 1992; 31:285-290.

190. Gelernter J, Goldman D, Risch N. The A1 allele at the D2 dopamine receptor gene and alcoholism. A reappraisal. J Am Med Assoc 1993; 269:1673-1677.

191. Noble EP, Blum K. Alcoholism and the D$_2$ dopamine receptor gene. J Am Med Assoc 1993; 270:1547.

192. Blum K, Noble EP, Sheridan PJ et al. Genetic predisposition in alcoholism: association of the D2 dopamine receptor TaqI B1 RFLP with severe alcoholics. Alcohol 1993; 10:59-67.

193. Pato CN, Macciardi F, Pato MT et al. Review of the putative association of dopamine D2 receptor and alcoholism: a meta-analysis. Am J Med Genet 1993; 48:78-82.

194. Gelernter J, Goldman D. Alcoholism and the D$_2$ dopamine receptor gene. J Am Med Assoc 1993; 270:1547-1548.

195. O'Dowd BF. Structures of dopamine receptors. J Neurochem 1993; 60:804-816.

196. Michel MC, Philipp T, Brodde O-E. α- and β-adrenoceptors in hypertension: molecular biology and pharmacological studies. Pharmacol Toxicol 1992; 70 (Suppl II): s1-s10.

197. Nijkamp FP, Engels F, Henricks PAJ et al. Mechanisms of β-adrenergic receptor regulation in lungs and its implications for physiological responses. Physiol Rev 1992; 72:323-367.

198. Simons FE. Evolution of H1-receptor antagonist treatment. Ann Allergy 1993; 71:282-287.

199. Van Tol HHM, Bunzow JR, Guan H-C et al. Cloning of the gene for a human dopamine D$_4$ receptor with high affinity for the antipsychotic clozapine. Nature 1991; 350:610-614.

200. Civelli O, Bunzow JR, Grandy DK. Molecular diversity of the dopamine receptors. Annu Rev Pharmacol Toxicol 1993; 32:281-307.

201. Coward DM. General pharmacology of clozapine. Br J Psychiatry 1992; 160 (Suppl

17):5-11.

202. Humphrey PPA, Fenuik W. Mode of action of the anti-migraine drug sumatriptan. Trends Pharmacol Sci 1991; 12:444-446.

203. Fenuik W, Humphrey PPA. The development of a highly selective 5-HT₁ receptor agonist, sumatriptan, for the treatment of migraine. Drug Devel Res 1992; 26:235-240.

204. Welch KMA. Drug therapy of migraine. New Engl J Med 1993; 11:1476-1483.

205. Kimball RW, A.P. F, Valejo E. Effects of serotonin in migraine patients. Neurology 1960; 10:107-111.

206. Lance JW, Anthony M, Hinterberger H. The control of cranial arteries by humoral mechanisms and its relation to the migraine syndrome. Headache 1967; 7:93-102.

207. Peroutka SJ, McCarthy BG. Sumatriptan (GR43175) interacts selectively with 5-HT₁ᴮ and 5-HT₁ᴰ binding sites. Eur J Pharmacol 1989; 163:133-136.

208. Van Wijngaarden I, Tulp MTLM, Soudjin W. The concept of selectivity in 5-HT receptor research. Eur J Pharmacol 1990; 188:301-312.

209. Feniuk W, Humphrey PPA, Perren MJ. CG43175 does not share the complex pharmacology of the ergots. Cephalalgia 1989; 9 (Suppl 9):35-39.

210. Collis MG, Hourani SMO. Adenosine receptor subtypes. Trends Pharmacol Sci 1993; 14:360-366.

211. Furlong T, Townsend-Nicholson A. Adenosine - More than fuel for thought. Today's Life Science 1992; 4:16-21.

212. Hökfelt T. Neuropeptides in perspective: the last ten years. Neuron 1991; 7:867-879.

213. Blundell J. Pharmacological approaches to appetite suppression. Trends Pharmacol Sci 1991; 12:147-157.

214. Leibowitz SF. Neurochemical-neuroendocrine systems in the brain controlling macronutrient intake and metabolism. Trends Neurosci 1992; 15:491-497.

215. Regoli D, Cadieux A, D'Orléans-Juste P. Vasoactive peptides and their receptors. Methods Neurosci 1993; 11:43-86.

216. Wahlestedt C, Reis DJ. Neuropeptide Y-related peptides and their receptors - are the receptors potential therapeutic drug targets? Annu Rev Pharmacol Toxicol 1993; 32:309-352.

217. McGillis JP, Mitsuhashi M, Payan DG. Immunomodulation by tachykinin neuropeptides. Ann NY Acad Sci 1990; 594:85-94.

218. Payan DG. Neuropeptides and inflammation: the role of substance P. Annu Rev Med 1989; 40:341-352.

219. Barnes PJ. Neurogenic inflammation and asthma. J Asthma 1992; 29:165-180.

220. Cooney M, Czernuszewicz G, Postel EH et al. Site-specific oligonucleotide binding represses transcription of the human c-myc gene in vitro. Science 1988; 241:456-459.

221. Holopainen I, Wojcik WJA. A specific antisense oligodeoxynucleotide to mRNAs encoding receptors with seven transmembrane spanning regions decreases muscarinic M₂ and GABAᴮ receptors in rat cerebellar granule cells. J Pharmacol Exp Ther 1993; 264:423-430.

222. Loke SL, Stein CA, Zhang XH et al. Characterization of oligonucleotide transport into living cells. Proc Natl Acad Sci USA 1989; 86:3474-3478.

223. Yakubov LA, Deeva EA, Zarytova VF et al. Mechanism of oligonucleotide uptake by cells: involvement of specific receptors? Proc Natl Acad Sci USA 1989; 86:6454-6458.

224. Baserga B, Denhardt DT, ed. Antisense strategies. Ann N Y Acad Sci. Vol 660. New York: 1992.

225. Gyurko R, Wielbo D, Phillips MI. Antisense inhibition of AT₁ receptor mRNA and angiotensinogen mRNA in the brain of spontaneously hypertensive rats reduces hypertension of neurogenic origin. Regul Pept 1993; 49:167-174.

226. Zhang M, Creese I. Antisense oligodeoxynucleotide reduces brain dopamine D₂ receptor: behavioral correlates. Neurosci Lett 1993; 161:223-226.

227. Wahlestedt C, Pich EM, Koob GF et al. Modulation of anxiety and neuropeptide Y-Y1 receptors by antisense oligodeoxynucleotides. Science 1992; 259:528-531.

# CHAPTER 5

# FUTURE PROSPECTS

Knowledge of the molecular architecture and the structural and functional diversity of the members of the G protein-coupled receptor superfamily is increasing at an exponential rate. An appreciation of the conservation of key structural components of these receptors, and the corresponding relationships at the DNA sequence level, has greatly facilitated the isolation of new members of this superfamily, including those encoding receptors which recognize as yet unidentified ligands. The molecular cloning of DNA sequences encoding these receptors has also facilitated the biosynthesis of large amounts of purified receptor for structural studies, as well as providing the basic templates for rapid analysis of structural determinants through in vitro mutagenesis and expression in heterologous systems. As approximately 80% of bioactive molecules, especially hormones and neurotransmitters, act through interaction with G protein-coupled receptors, it is not surprising that this rapidly increasing knowledge of the detailed structure and function of this receptor class is providing growing insight into the molecular pathology of a wide range of disorders, as well as providing important new tools for the development of novel receptor subtype specific therapeutic agents.

Although manipulation of the activity of members of this receptor class will undoubtedly be used across the whole field of biomedical research, it is already clear that one of the most exciting areas of the future will be analysis of the role of these receptors in the regulation of brain function. A major aim of molecular neurobiology is to understand how a very few neurotransmitters and neuropeptides can elicit such a wide variety of neuronal responses. The G protein-coupled receptors and the diverse signaling mechanisms to which they couple are intimately linked to the maintenance of normal neuronal function. Imbalances in some of these G protein-coupled chemical signaling pathways in the nervous system are prime candidates for a role in the development of mental illness, just as their normal functioning is most probably involved in a wide range of inherent and acquired behavioral responses.

## STRUCTURAL CONSERVATION
## AND FUNCTIONAL DIVERSITY

Beginning with the initial observation that the structure of the adrenergic receptors was similar to that of bacteriorhodopsin, the molecular cloning of DNA sequences encoding several hundred G protein-coupled receptors has very clearly demonstrated that the paradigm of a seven transmembrane polypeptide chain coupled to a complex of G proteins is an ancient theme which has been copied innumerable times. It is now clear that hundreds of variations on the same theme play a central role in the signaling pathways of bioactive molecules as diverse as small catecholamines, e.g. noradrenaline, large glycosylated dimeric polypeptide hormones such as FSH, bioactive peptides, inflammatory mediators and photons of light. Furthermore, these variations on the common theme have resulted in a wide range of novel adaptations to achieve receptor activation. These range from intimate binding in the transmembrane pocket, as seen with the catecholamines, binding to an extended N-terminal region as with FSH, enzymatic activation (e.g. thrombin) to incorporation of a photo-activatable chemical, as found in rhodopsin. As even more G protein-coupled receptors are isolated by molecular cloning and the detailed basis of ligand binding and receptor activation elucidated, there will undoubtedly be many other novel mechanisms identified which are used to transduce ligand binding into receptor activation. The detailed molecular basis of this signal transduction event remains one of the major goals of G protein-coupled receptor research. Elucidation of the presumed subtle conformational changes elicited by ligand binding will require innovative application of a range of structural techniques.

To date, most of the activity in the molecular biology of G protein-coupled receptors has focused on the initial cloning of the receptor encoding DNA sequences, expression of the receptor in various heterologous systems and analysis of the specificity and complexity of second messenger coupling. However, the availability of cloned receptor sequences is also providing the templates for detailed studies of structure-function relationships using in vitro mutagenesis. To date, these have tended to focus on the adrenergic receptor family, but are rapidly being extended to the other receptor subfamilies. These studies will be greatly facilitated in the future

once the complete group of receptor subtypes for any particular ligand has been isolated. In such cases, a comparison of conserved and variable amino acid residues, and the ability to create chimeric but functional receptors, often provides a focus to more directed in vitro mutagenesis studies. As such studies on a range of receptor subfamilies are undertaken, significant insight will be achieved into aspects of G protein-coupled receptor structure which are common to the whole superfamily as well as to those which are important only to the specific receptor subclasses. Although, to date, in vitro mutagenesis has tended to focus on identification of determinants important for agonist and antagonist binding, similar approaches offer enormous potential as a means to identify the basis of receptor-G protein coupling.

As new G protein receptor subtypes are being isolated by molecular cloning, mutations are being identified in many subtypes and shown to be linked to a wide range of human diseases. Such "natural" mutations will be uncovered at an increasing rate in the future and will complement the site directed mutagenesis studies in providing more insight into the mechanism of action of this receptor family. For example, a surprising number of constitutively active receptor mutations have already been identified in several different receptors. The location and nature of these mutations are extremely varied and suggest that in general this class of receptor is normally in a state of inhibition, with ligand binding causing a release of an inhibitory constraint. As the level of analysis of these receptors extends to investigation of polymorphisms in the population, many nonpathological variants will undoubtedly be found which contribute to the extensive natural variation seen in a whole range of physiological parameters. Evaluation of such natural polymorphisms and their functional consequences will also contribute to our understanding of receptor structure and function at an even more detailed level.

The functional diversity generated by the existence of multiple receptor subtypes

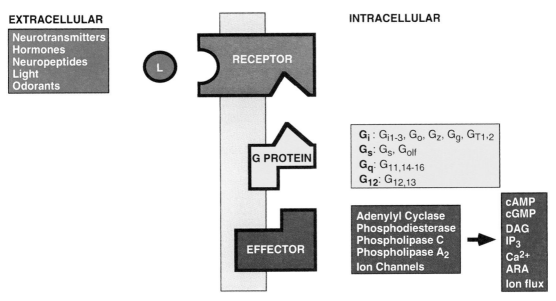

Fig. 5.1. Multiplicity of components in G protein-coupled receptor signaling pathways, incorporating a variety of ligand (L), G proteins and effector systems.

for any given ligand has been shown to be due to both the generation of unique gene products, presumably arising from gene duplication events, as well as from alternative splicing events following transcription from a single gene. Such fairly dramatic structural differences between receptor subtypes have been relatively easy to identify and correlate with functional differences. However, there will undoubtedly be more subtle, often regulatory, differences identified in the future. Mechanisms such as alternative splicing of different 5' untranslated regions, transcription from different promoters in different cell types and other means of transcriptional and/or translational control will be increasingly found as the biosynthesis of specific receptor subfamilies is examined in more detail.

One of the most important outcomes of the molecular cloning and expression of this wide variety of G protein-coupled receptors has been the demonstration that specific receptor subtypes couple to very specific G proteins and effector systems and, furthermore, that such coupling is not absolute, but very dependent upon the specific repertoire of G proteins and effector systems available in any particular cell

type. Enormous progress is currently being made in understanding which particular signal transduction pathway is important in which particular physiological response. This progress is being dramatically facilitated by the similar isolation, through molecular cloning, of a growing family of G protein subunits and an increasing array of molecules involved in different effector systems as illustrated in Figure 5.1. As further members of these gene families are isolated and characterized, our understanding of the unique composition of specific G protein-coupled receptor pathways will be greatly enhanced.

The use of the substantial homology at the DNA sequence level between different members of the G protein-coupled receptor superfamily is also providing novel approaches to the identification and isolation of new bioactive ligands. This progress is occurring on two fronts. The first involves the isolation of endogenous ligands for known receptors previously identified by their affinity for various exogenous substances, e.g. isolation of anandamide as the naturally occurring ligand for the cannabinoid receptor. On the other hand, a wide range of receptor molecules have been isolated for which there is currently no known ligand. These so called "orphan" receptors may well provide the tools to isolate previously unidentified neuropeptides or neurotransmitters and other biologically important ligands.

## MOLECULAR PATHOLOGY

The increasing number of G protein-coupled receptors implicated as the molecular defect in a range of pathological conditions is to be expected given the widespread distribution of these receptors, the diversity of ligands with which they interact and their central role in signal transduction mechanisms. Although the identification of recessive conditions resulting from loss of function mutations in different G protein-coupled receptors may have been expected (e.g. color blindness) and undoubtedly many more will be found, the already relatively large number of mutations which have been identified that cause constitutive activation of receptors is somewhat surprising. Despite the known presence of a variety of mechanisms to down-regulate activated G protein-coupled receptors, ranging from desensitization to transcriptional modification, such constitutive mutations often clearly escape these feedback mechanisms. Already mutations that cause constitutive signaling by G protein-coupled receptors have been found to occur at a wide variety of different sites in the receptors, suggesting that many more disorders due to constitutive mutations in this class of receptor will be found in the future. Given the important and widespread role of this class of receptors in the central nervous system, it would be expected that polymorphisms in many of these receptors may play some role in the generation of differences (both substantial and subtle) in various neuronal pathways. As different receptor subtypes couple to different effector systems depending on the availability of the components of such systems, receptor mutations or polymorphisms will undoubtedly be also shown to result in a shift in the balance between coupling to various signal transduction pathways

with a subsequent shift in the homeostasis of the cell.

## NEW DRUG DEVELOPMENT

The isolation of the various members of the G protein-coupled receptor superfamily provides important tools for the development of highly selective therapies and new preventative approaches to many complex multifactorial diseases. These opportunities arise from a detailed understanding of the structure and function of individual receptor subtypes in different disorders and have dramatically changed both the concepts and techniques of pharmaceutical drug development. Over the past few decades, sophisticated organic chemistry and pharmacology has been used to modify small molecule ligands such as noradrenaline and histamine to provide new selective agonists and antagonists with restricted therapeutic activity and limited side effects. The use of such specific analogs has illustrated the presence of structurally different subtypes of receptor, which often couple to different second messenger pathways and hence different functions. An understanding of the role of different receptor subtypes in different physiological pathways (e.g. $\alpha$- and $\beta$- adrenergic receptors) has provided the basis for development of subtype-specific drugs to treat a range of complex disorders. Until recently this approach has been limited to small molecule ligands. Knowledge of the range of receptor subtypes used by the larger peptide and protein ligands has been restricted because of the difficulty in developing selective agonists and antagonists of these much larger and more complex molecules. However, molecular cloning of the receptor subtypes has now resulted in a similar capability for receptors recognizing the larger peptide and protein ligands. Furthermore, a detailed knowledge of the structure-function relationships of the different receptor subtypes provides novel opportunities for the development of more selective agonists and antagonists. It is important to remember that the availability of novel and highly specific molecular targets through the cloning of G protein receptor subtypes is paralleled by similar exciting developments in new approaches to antisense and gene therapy as well as rapid progress in the molecular modeling of both small and large ligands. Together, these technological developments are changing the approach to development of new therapeutic substances. The specificity and sensitivity of G protein-coupled receptors may also be used in other non pharmaceutical applications such as development of biosensors using odorant receptors or receptor probes to monitor levels of drug treatment.

As the range of ligands acting through G protein-coupled receptors modulates many important homeostatic pathways, the development of new pharmaceuticals based upon our knowledge of the existence of receptor subtypes has fallen into three major areas. Initially, this has involved mainly the development of improved drugs acting on existing target pathways, e.g. the renin-angiotensin system, but is rapidly being extended to the development of new classes of drugs acting at pathways not previously targeted, for example many of the neuropeptides and neuromodulators. Furthermore, interest is rapidly growing in the opportunity to develop drugs targeted to novel previously unidentified pathways and previously unidentified ligands, an opportunity which has arisen from isolation of "orphan" receptors.

## CONCLUSION

Because of their widespread distribution and importance in a range of physiological pathways, increasing insight into the structure and function of G protein-coupled receptors will have ramifications in virtually all fields of biomedical research. However, because of the importance of this receptor family in modulating and integrating the complexity of chemical signaling in the nervous system, such insight is likely to be of particular significance in elucidating the molecular basis of neuronal function and the pathophysiology of mental illness. At the same time, the availability of cloned receptor subtypes will herald a new era of rational drug design and screening to provide novel therapies to address the complex multifactorial diseases challenging modern medicine.

# INDEX